Physical Chemistry of Macromolecules

Gary Patterson

Department of Chemistry
Carnegie Mellon University

CRC Press
Taylor & Francis Group
Boca Raton London New York

CRC Press is an imprint of the
Taylor & Francis Group, an **informa** business

CRC Press
Taylor & Francis Group
6000 Broken Sound Parkway NW, Suite 300
Boca Raton, FL 33487-2742

First issued in paperback 2019

© 2007 by Taylor & Francis Group, LLC
CRC Press is an imprint of Taylor & Francis Group, an Informa business

No claim to original U.S. Government works

ISBN-13: 978-0-8247-9467-5 (hbk)
ISBN-13: 978-0-367-38938-3 (pbk)

Library of Congress Cataloging-in-Publication Data

Patterson, Gary B.
    Physical chemistry of macromolecules / Gary Patterson. -- 2nd ed.
        p. cm.
    Includes bibliographical references and index.
    ISBN-13: 978-0-8247-9467-5 (alk. paper)
    ISBN-10: 0-8247-9467-2 (alk. paper)
    1. Polymers. I. Title.

QD381.7.P38 2007
547'.7045--dc22   2006030992

Visit the Taylor & Francis Web site at
http://www.taylorandfrancis.com

and the CRC Press Web site at
http://www.crcpress.com

Contents

Preface

This book is the culmination of 40 years of thinking about macromolecules. It began when I considered where to attend graduate school and read Paul Flory's[15] *Principles of Polymer Chemistry* for the first time. It continued at Stanford University as a member of the Flory laboratory. The application of statistical mechanics to the understanding of the properties of chain molecules was pursued during these years. During this time period I made the acquaintance of many polymer scientists. I learned from Robert Pecora that light scattering is a useful and informative technique for understanding molecular systems. I was also privileged to meet Eugene Helfand while he was a visitor to Stanford in the Flory laboratory for a year. Our discussions continued at AT&T Bell Laboratories during the years from 1972 to 1984.

The polymer science community at Bell Labs was large and diverse. William O. Baker, president, would often drop by to discuss the details of my studies of the structure and dynamics of liquid polymers. David W. McCall, director, was a constant source of advice and encouragement. Shiro Matsuoka endeavored to teach me the importance of chemical engineering for the understanding of polymeric materials. Frank A. Bovey and Field ("Stretch") Winslow, editors of *Macromolecules*, opened up the full range of polymer science on a daily basis. Virtually every major figure in polymer science visited Bell Labs during my time there (1972 to 1984), and the frequent opportunities for travel and lecturing led to personal knowledge of most of the key contributors to the physical chemistry of macromolecules. In addition, the large number of colleagues in polymer science provided an environment where any idea could be discussed in detail in rapid order. Good ideas could be critiqued and bad ideas could be demolished.

Carnegie Mellon University is one of the important cradles of polymer science in the U.S. Edward Cassasa, editor of the *Journal of Polymer Science*, provided a vital link to the historical richness of the science of macromolecules. Guy Berry was an invaluable resource to everything macromolecular. The opportunity to teach courses in the physical chemistry of macromolecules provided the motivation to create a coherent presentation of the principles of the discipline.

This book was developed as a series of lectures for a graduate course in the physical chemistry of polymers. It assumes an advanced undergraduate knowledge of physical chemistry. It contains much more than can be covered in one semester, but it is neither comprehensive nor exhaustive. It focuses on a coherent picture of polymer science inspired by pioneers like Flory, deGennes, and Ferry. Topics are developed in enough detail to inspire confidence in the results, and the limitations of the treatments are discussed explicitly. No attempt has been made to present the most detailed level of understanding available in the research literature. The goal of the lectures was to make it possible to continue the development of the student to the point where the classic books and papers of the science of macromolecules could be read and understood. Nevertheless, the eventual level of the book will be a challenge to students (or scholars) with a limited knowledge of physical chemistry or polymer science. It is hoped that this book will be helpful to anyone trying to understand the science of macromolecules at a level high enough to work in this discipline.

The book opens with a historical reflection on the emergence of a scientific discipline known as polymer science. The influence of Herbert Morawetz on my own thinking in this area is significant. In addition, a historical project on "Paradigms in Polymer Science" is currently in progress at the Chemical Heritage Foundation. I am grateful for the award of the Charles C. Price Fellowship in Polymer History that facilitated that work and resulted in the completion of the present volume.

The concrete visualization of macromolecules necessary to formulate detailed predictions of properties provides the fundamental conceptual world that defines this book. Macromolecules are constrained by the same forces that define any molecule, and the atomic level of description is essential for many problems. The inspiration for the presentation found here for the rotational isomeric state model is the classic book by Flory,[19] *Statistical Mechanics of Chain Molecules*. When more-global properties are considered, smoothed models such as the Gaussian chain or wormlike chain model are more tractable. The importance of selecting a model that is both empirically adequate and mathematically solvable is stressed.

While this book is primarily theoretical, and the number of equations is very large, the experimental basis of polymer science is always in view. Three experimental techniques that have been important in the development of the understanding of the macromolecular paradigm are featured in Chapter 3. The ability of nuclear magnetic resonance (NMR) to distinguish the local chemical environment and the time scale for molecular motion has made this a valuable tool in the elucidation of polymer structure and dynamics. Many of these insights are presented in a monograph by Frank A. Bovey.[4] The experimental basis for the rotational isomeric state model is often neglected. Actual rotational isomeric states can be experimentally visualized using vibrational spectroscopy, and R.G. Snyder[38] has demonstrated this in detail. Another of the key experimental techniques that have contributed to our current paradigm for chain molecules is light scattering. The historical development of this topic is rich with insights into the nature of macromolecules. With the invention of the laser and the publication of the classic monograph of Berne and Pecora,[3] light scattering has been established as a central tool in polymer science.

The phenomenon of rubber elasticity fascinated scientists of all kinds during the 19th century. The explanation of the temperature dependence of the force of retraction in terms of the Gaussian theory of chain statistics was one of the first great triumphs of the macromolecular paradigm. One of the recurrent themes of this book is the clear explanation of actual phenomena in terms of the macromolecular nature of the system.

The existence of dilute solutions of macromolecules was denied by many experts until the macromolecular hypothesis was largely accepted in the time period from 1930 to 1940. The dilute-solution state is still the basis for characterizing individual macromolecules and the interactions of pairs of macromolecules and the solvent. The structural, thermodynamic, and hydrodynamic properties of polymer solutions are explained in terms of the random-coil model developed by Kuhn, Debye, Flory, Kirkwood, Yamakawa, and deGennes. While this subject alone could easily be the basis for a one-semester course, the topics are developed so that the material could be presented as part of a complete development of the subject.

Individual random-coil chain molecules pervade a very large volume compared with the actual chemical volume of the chain. The conceptual framework for understanding the properties of solutions concentrated enough so that the pervaded volume of the chains is comparable to the total solution volume — the so-called semidilute state — has been presented in the classic book by Pierre-Gilles deGennes,[9] *Scaling Concepts in Polymer Physics*. In addition to the discussion of the coil size, osmotic pressure, and mutual-diffusion coefficient of chains in semidilute solution, new results are given for the viscosity and structure of these solutions. Light-scattering studies reveal liquidlike structure for the coils in semidilute solution.

The thermodynamic properties of concentrated polymer solutions were studied by Flory[15] and independently by Huggins.[21] The Flory–Huggins theory of polymer solutions still forms the basis for much discussion of these solutions in industry and even in academic research. Understanding this model is important for making connections to much of the literature. Flory also substantially improved this model to include compressible fluids. The Flory–Orwoll[33] theory of polymer solutions is still transparent and easily applicable, predicting both upper and lower critical solution temperatures. More-empirically adequate theories of concentrated solutions do not lend themselves to simple lecture presentation and often require detailed computer calculations to obtain any results. Concentrated solutions also introduce the phenomenon of viscoelasticity. An extensive treatment of the full distribution of relaxation times necessary to understand the dynamic properties of polymers in concentrated solution is presented.

While there are many sophisticated theories of the equation of state for amorphous polymer liquids, a simple free-volume equation of state has been found to be useful for many problems. The viscosity of bulk amorphous macromolecules is developed and discussed as a function of molecular weight, temperature, and pressure. The properties of liquid polymers are explained in terms of a distribution of relaxation times, and many experimental techniques that manifest this phenomenon are discussed. One of the most important observable properties of amorphous liquid macromolecules is the glass transition. The paradigmatic observations that define the glass-transition region and the explanation in terms of the path-dependent state of the material are presented.

Not all chain molecules are random coils, and the properties of rodlike molecules in solution are of great current interest. The characterization of stiff chains in dilute solution using light scattering is discussed. The phase behavior has been considered by Flory,[17] and the statistical thermodynamics of such systems is presented. The appearance of a miscibility gap between isotropic solutions and lyotropic solutions is derived. Inclusion of intermolecular attractive forces leads to further phase separation into very dilute and very concentrated solutions.

Many macromolecules in aqueous solution are polyelectrolytes. The remarkable changes in the conformation of linear polyelectrolytes as a function of concentration, ionic strength, and pH are discussed. The various theories of chain expansion are reviewed. The thermodynamic properties of polyelectrolyte solutions reveal dramatic behavior. The large increase in the reduced osmotic pressure, π/c, as the solution is diluted is explained in terms of the entropy of the counterions. The strong dependence of the conformation of the chains with solution conditions also leads to large changes in the viscosity. The viscosity is also explained in terms of the coil size and the interactions of the chains.

Highly ordered systems such as crystalline polymers or microphase-separated block copolymers are not treated in this volume. There are excellent books that treat these highly visually oriented topics in great detail. While there are enough figures in this book to illustrate the main experimental results on which the macromolecular paradigm rests, the book is not graphically intensive.

The Polymer Pioneers:

This group photo was taken in 1981 at the Plaza Hotel, New York, on the occasion of the party held for Herman Mark's 85th birthday. It includes many of the time's leading figures in polymer science. A partial list of attendees is: *Back row:* Ranby, Unknown, Ringsdorf, Morawetz, Ogata, Bikales, Winslow, Merrifield, Goodman. *Second row:* Overberger, Morgan, Pino, Pearce, Fields, Smets, Eirich, Unknown, Marvel. *Front row:* Unknown, Saegusa, Pauling, Vogl, Herman Mark, Hans Mark, Karchalski-Katzir, Flory, Unknown. From the Carl Marvel Archives, Chemical Heritage Foundation, with permission.

Acknowledgments

The community of polymer scientists is now large and very diverse. The discipline is inherently multidisciplinary and requires input from many perspectives. I wish to thank E.L. Thomas (MIT, Material Science and Engineering Department) for insights into the structure of polymeric materials and for Figure 1.2. Frank Bovey and Lynn Jelinski helped with the section on NMR (Section 3.3). Robert G. Snyder provided detailed advice on vibrational spectroscopy of polymers and Figure 3.3. Ben Chu has exemplified the highest standards in light scattering and produced Figure 5.1.

I wish to thank my many collaborators and students, especially James R. Stevens, Patrick Carroll, and S.H. Kim. Careful experimental work has firmly established the phenomenological basis of polymer science.

Finally, I wish to thank my wife, Susan Patterson, for joining me in 40 years of polymer musings.

About the author

Gary Patterson is a chemical physicist with interests in polymer science, complex fluids, and colloid science. He attended Harvey Mudd College (B.S., chemistry, 1968) and Stanford University (Ph.D., physical chemistry, 1972). He was a member of the technical staff in the Chemical Physics Department at AT&T Bell Laboratories from 1972 to 1984. He is a Fellow of the Royal Society of Chemistry and of the American Physical Society. He also received the National Academy of Sciences Award for Initiatives in Research in 1981. He has been a professor of chemical physics and polymer science at Carnegie Mellon University since 1984. In addition to teaching physical chemistry to chemical engineers and chemists, he teaches in the College of Humanities and Social Sciences. He conducts an active research program in experimental and theoretical chemical physics, with emphasis on the structure and dynamics of macromolecular systems.

chapter one

Introduction to macromolecules

1.1 What is a macromolecule?

To define a macromolecule, it is necessary first to consider the basic definition of a molecule. A molecule is a group of atoms connected by covalent chemical bonds. The chemical structure of the molecule is defined by the specific connectivities between the atoms. Some atoms can covalently bond to more than one other atom. The study of macromolecules explores the rich variety of chemical structures that can be created by assemblies of atoms that display multiple valence.

The detailed chemical structure of a macromolecule is defined by the location of all the atoms and by the specification of the pairs of atoms that are covalently bonded. A full discussion of this subject is presented in Chapter 2. It is often convenient to divide the structure of a macromolecule into subunits called **mers**. A molecule that can be divided into a small number of subunits is called an **oligomer**. If the number of subunits is large the molecule is called a **polymer**.

Macromolecules are classified by considering the number and types of subunits and by the topology of the structure (Figure 1.1). Some molecules are composed of a closed linear array of subunits and have no ends. If all the subunits are identical, the molecule is called a **homopolymer**. Linear molecules with end subunits are also called homopolymers if all the interior subunits are identical, but it must be remembered that the properties of the molecule may depend strongly on the end units if the number of mers is small. Molecules with more than one type of interior subunit are called **copolymers**. If any of the subunits are connected to more than two other subunits, the structure is called **branched**. If two branch points are connected by more than one continuous path of mers, the structure becomes a **network**. The rich variety of topologies that can be created by subunits with functionalities greater than two is one of the wonders of polymer science.

POLYMER TOPOLOGIES

Homopolymer: E (M)$_n$ E'

Copolymer: E M$_1$M$_2$M$_3$...M$_i$ M$_{i+1}$ ...M$_n$ E'

Branched: E M$_1$...B$_i$ M$_{i+1}$ ...M$_n$ E'
$$M_j$$
$$\vdots$$
$$M_k$$
$$E''$$

Network: E M$_1$ B$_1$ M$_i$...M$_j$ B$_2$...M$_p$ E'
                       M$_k$ M$_n$
                        $\vdots$ $\vdots$
                       M$_1$.................M$_m$

Figure 1.1 Polymer molecules can be classified as homopolymers or copolymers. The overall topology can be linear, branched, or some form of a network.

1.2   The discovery of macromolecules

The discovery of molecules was greatly aided by the realization of Avogadro that the pressure, P, of a dilute gas at a known temperature, T, and volume, V, was determined by the number of particles, N, independent of the mass of the molecules. Measurement of the gas density, ρ, then allowed a calculation of the molecular weight, M:

$$M = N_A \text{ (mass of gas/number of molecules)} \tag{1.1}$$

where $N_A = 6.022 \times 10^{23}$ mol^{-1} is Avogadro's number. It was also possible in the early 19th century to measure the atomic composition of many substances. Faraday used the gas density method to discover a molecule with the atomic composition of ethylene (C_2H_4) but with twice the molecular weight. He discovered the molecule butene. Gas density can be used to study a wide range of oligomers, but as the molecular weight increases, the equilibrium vapor pressure decreases, and high polymers cannot be studied at equilibrium in the gas phase.

Many macromolecules can be dissolved in an appropriate solvent to produce a dilute solution. The dilute solution state has served as the basis for many of the fundamental conceptual advances in polymer science. Raoult established the quantitative relationship between the melting-point depression of a solvent and the number density of the solute. The same procedure developed to measure molecular weight in the dilute gas phase could then be used in dilute solution. One of the first macromolecules to be studied in this way was starch, which was shown to have a molecular weight near 30,000 g/mol. Such a high molecular weight was rejected by many authorities, and cryoscopic determinations of molecular weight were questioned. Starch can be converted to a smaller macromolecule called dextrin. The molecular weight of dextrin was determined by another important technique: end-group analysis. In this case, the end group was converted chemically to a carboxylic acid, and the number of end groups in a sample of known mass was measured by titration. The molecular weight of dextrin was shown to be the same from both cryoscopic measurements and from end-group analysis. This historical account illustrates one of the most important principles of polymer science: **physical chemistry is fully applicable to macromolecules**.

Another important experimental technique that has molded the thinking of polymer scientists is osmometry. One of the most brilliant polymer scientists, van't Hoff, showed that the osmotic pressure of a dilute solution is determined by the temperature and the number density of solute molecules, in complete analogy to an ideal gas. Many proteins were then shown to be macromolecules with molecular weights in excess of 10,000 g/mol. Again, many leading physical chemists insisted that particles with high molecular weight in solution were colloidal in nature and consisted of noncovalent assemblies of small molecules. Hermann Staudinger received the Nobel Prize in chemistry for his tireless efforts to establish the concept of macromolecules. **Covalent molecules are the basis of polymer science**.

1.3   The structure of macromolecules in solution

The detailed chemical structure of many small molecules is dominated by intramolecular interactions, and the shape is independent of the state of macroscopic aggregation (gas, liquid, or solid). On the other hand, the shape of a macromolecule is often a function of its environment. One of the earliest evidences of this fact was the large difference in the intrinsic viscosity, $[\eta]$, of macromolecules in different solvents. The measurement of solution viscosity has played a central role in the development of polymer science. The theory of hydrodynamics was well developed by the end of the 19th century, and Einstein derived the equation for the viscosity of a dilute solution of rigid spheres with volume fraction φ:

$$\eta = \eta_0\left[1 + (5/2)\varphi\right] \qquad (1.2)$$

where η_0 is the viscosity of the pure solvent. A knowledge of the number density of solute molecules then allows the determination of the hydrodynamic volume, $V_h = \varphi/(N_2/V)$, of the macromolecules. For globular protein molecules, the hydrodynamic volume is well accounted for by the sum of the volumes of the constituent peptide subunits.

The intrinsic viscosity is defined as:

$$[\eta] = \lim_{c \to 0} \frac{\eta - \eta_0}{c\eta_0} \qquad (1.3)$$

where $c = (N_2 M/N_A V)$ is the mass concentration of the solute. For a solution of particles with hydrodynamic volume V_h, the intrinsic viscosity is given by:

$$[\eta] = [(5/2)V_h N_A]/M . \qquad (1.4)$$

The intrinsic viscosity of globular proteins in a suitable buffer solution is independent of molecular weight! This conclusion illustrates another important principle of polymer science: **proper interpretation of experimental data requires a valid theory.**

Studies of soluble homopolymers in thermodynamically good solvents revealed that the intrinsic viscosity increased with molecular weight. Staudinger attempted to summarize these results with the empirical relation:

$$[\eta] = KM \qquad (1.5)$$

where K is a constant that depends on the solvent, the temperature, and the specific macromolecule. He then proceeded to rationalize this apparently universal law by claiming that it was best explained by a rigid linear structure for the polymer. This procedure illustrates one of the historical pitfalls that have plagued polymer science: **bad experimental results that are justified by bad theory.**

A more careful examination of the experimental data by Mark led to a more general relationship between the intrinsic viscosity of macromolecules in solution and molecular weight:

$$[\eta] = KM^a \qquad (1.6)$$

where the exponent a also depends on the solvent, the temperature, and the specific macromolecule. He also pointed out that, for rigid rods, the intrinsic viscosity actually depends on M^2. Many polymers observed in thermodynamically good solvents yield exponents in the range 0.6 to 0.8.

The correct explanation of this result by Kuhn is one of the first triumphs of the statistical theory of polymer chains. He was well aware of the developments in structural chemistry that explained the flexibility of molecules in terms of rotation about single bonds. Since polymers are molecules, rotation about the bonds along the main chain of the macromolecule would lead to a very large number of conformations. He was able to show that the average hydrodynamic volume of such a system would increase as $M^{3/2}$. This transparent result was still not in agreement with the observed result. Kuhn then appealed to the finite physical volume of each subunit and claimed that the chain would expand further due to the "excluded volume." This effect then yielded an intrinsic viscosity that depended on molecular weight with an exponent in the proper range, but the details of the correction were not given. A quantitative theory of the effect of excluded volume on the structure of statistical polymer chains in solution was eventually presented by Flory.[14]

If the refractive index of the solution depends on the concentration of the solute, the motion of the solute can be followed by optical means. The earliest use of this principle was to observe pollen particles diffusing in water, demonstrating the phenomenon of Brownian motion. Einstein correctly explained this process as due to the random kinetic motions of the solvent molecules. The solute was characterized by a friction coefficient, $f = 6\pi\eta_0 R_h$, where R_h is the hydrodynamic radius of the particle. The diffusion coefficient of the particle in the solution was shown to be $D = k_b T/f$, where k_b is the Boltzmann constant.

Another consequence of the treatment of dilute solute molecules as statistical particles is the conclusion that the equilibrium concentration of solute particles in a gravitational field should vary with the height in the solution. The distribution function derived by Einstein is:

$$c(h) = c(0)\exp[-Mg(1-\bar{v}\rho)h \,/\, RT] \tag{1.7}$$

where g is the gravitation constant, $\bar{v}$ is the partial specific volume of the solute, and $R = k_b N_A$ is the gas constant. Perrin received the Nobel Prize for experimentally demonstrating that this distribution function was valid for suspended particles of any type, including macromolecules.

For typical macromolecules, diffusion over macroscopic distances is a very slow process. The settling rate of the same macromolecules in a gravitational field is also a slow process. To speed up the process, Svedberg invented the ultracentrifuge. The sedimentation constant s for a solute particle is given by:

$$s = M(1-\bar{v}\rho) \,/\, N_A f \;. \tag{1.8}$$

If the friction coefficient can be calculated for the solute particle, or if the diffusion coefficient can be measured independently, the sedimentation constant can be used to determine the molecular weight of the solute particle. Svedberg determined the molecular weights for many proteins using this technique and earned the Nobel Prize for his work. The molecular weight can also be determined from the distribution of particle concentration at sedimentation equilibrium.

Although hydrodynamic measurements provided important information about the structure of macromolecules in dilute solution, the direct relationship between hydrodynamic radius or volume and the geometric structure of the molecule was still not known. A more direct measure of the spatial structure of dissolved macromolecules was obtained with light scattering. Debye derived the formula for the intensity of scattered light from a solute as a function of scattering angle. The geometric quantity obtained for a pure homopolymer is the mean-squared radius of gyration $\langle R_G^2 \rangle$. Light scattering can also be used to measure the molecular weight of solute particles. The measured relationship between the molecular weight, M, and $\langle R_G^2 \rangle$ has inspired many advances in the understanding of macromolecules.

1.4 The remarkable properties of pure bulk polymers

Macromolecules can exist in pure condensed states. One of the most remarkable natural substances is Hevea rubber, obtained from the rubber tree (*Hevea brasiliensis*). Samples of this material were brought to Europe soon after Columbus made contact with South America. Faraday purified it sufficiently to obtain an elemental analysis that showed it was a pure hydrocarbon.

The most well-known property of a rubber band is its elasticity: the material can be stretched to many times its original length and then return to its original length when the tension is released. Another startling property is observed when a rubber band stretched by a suspended weight is heated: the rubber band shortens as the material is heated. These phenomena interested some of the best theorists of the 19th century (Kelvin) and the best experimentalists (Joule). The proper explanation of the phenomenon of rubber elasticity required the development of the statistical theory of polymer chains by Kuhn. The further work of Guth and Mark firmly established that the retractive force was due to the reduction in the entropy of the system due to the restriction on the conformations of the chains brought about by stretching the material. The success of the statistical theory of rubber elasticity implied that the conformations of polymer chains in pure bulk materials were similar to those in dilute solution. The unity of polymer science is based on the behavior of macromolecules. It is now known that any macromolecule that is observed above the melting point of its crystalline phase, and for which there are many available conformations, will exhibit rubber elasticity if the chains are long enough.

When rubbery materials are cooled, two typical outcomes are obtained: (1) the material partially crystallizes to become a flexible semicrystalline substance or (2) the material does not crystallize and, at sufficiently low temperature, becomes an amorphous glassy solid. One of the most confusing aspects of the behavior of macromolecules during the 19th century was the observation of crystallization. It was believed that polymers could not crystallize because the long chains would never achieve the fully regular structure needed for crystallization unless they were already fixed in such a structure. The actual observation of crystalline polymers convinced Staudinger that the only valid conformation for a chain molecule was a rigid helix. If the macromolecule was observed in the pure bulk state above the melting point, rigid rods would not yield a rubbery isotropic material, but would instead be in the liquid crystalline state. Another quandary was the observation of unit cells in crystalline polymers that were the size of only a few mers. The leading X-ray spectroscopists insisted that a unit cell had to be larger than the size of the molecules of which it was composed. The resolution of this problem by Mark was actually fully anticipated by the brilliant work of Polanyi, more than a decade earlier! The unit cell in a crystalline polymer sample contains mers from a small group of helical chains determined by the space group of the crystal. The repeat units of the chain then replicate the unit cell. X-ray scattering can also be used to determine the size of the crystals in a polycrystalline sample. The individual crystallites in most semicrystalline polymer samples have dimensions that are small with respect to the fully extended length of the individual macromolecules. The flexibility of such materials is explained by the amorphous rubbery behavior of the intercrystalline material.

A long controversy arose over the extent to which any particular chain was confined to a single microcrystal. It is now understood that one chain can be part of many microcrystals and that the tensile modulus of a semicrystalline polymer sample is related to the modulus of the tie chains that connect individual microcrystals. The number of different space groups observed in crystalline polymers and the overwhelming number of different morphologies that are found are a reminder that polymer science is an unending frontier of research.

The existence of pure amorphous bulk polymers has been a controversial issue since the beginning of polymer science. Natural rubber yields an X-ray pattern that contains only amorphous halos, typical of any liquid. Nevertheless, it was difficult for many scientists to believe that molecules with a polymeric chain structure could pack in a truly amorphous way. There are still papers that are submitted for publication that assert that amorphous rubbery polymers are actually composed primarily of microcrystalline domains. This issue has been clarified by the incisive theoretical and experimental work of Flory. It is now understood that there are polymers that exhibit liquid crystalline phases upon melting of the crystals. The nature of the noncrystalline state of pure bulk polymers depends on the detailed local structure of the chain and the ratio of the persistence length of the chain to the diameter of the mer. Molecules that are conformationally flexible enough to have a small persistence length can exist in the amorphous liquid state.

There are many macromolecules that are amorphous liquids at high temperature and that do not have a crystalline phase. These systems are actually copolymers of various types. When these liquids are cooled, their viscosity increases enormously. Eventually they exhibit shear moduli in the solid range as long as the measurements are carried out at a finite frequency or over a short period of time. Many polymers are used commercially in the amorphous glassy state. Even macromolecules that do have known crystalline phases can be prepared in the glassy state by rapid quenching to avoid crystallization. Liquid crystalline polymers can also be prepared in a glassy state by rapid quenching to yield an ordered solid.

1.5   *Building new materials using macromolecules*

The properties of mixtures of pure substances can be very different from the parent materials. The art of metallic alloys is an important part of industrial practice. While there are a few polymer blends that are truly compatible in a thermodynamic sense, most pairs of pure homopolymers do not mix. The theoretical reason for the low tendency to mix is the very small entropy of mixing of macromolecules. The entropy of mixing depends on the number density of molecules, which will be small if each molecule occupies a large volume.

Another approach to creating materials with properties different from any pure homopolymer is to use copolymers. Since the different mers are covalently attached to the macromolecule, they cannot undergo macroscopic phase segregation. The local structure of a bulk copolymer depends on the sequence distribution of the constituent subunits. Consider a copolymer composed of two types of subunits, a so-called AB copolymer. The total number of mers, N, is the sum of type A subunits, N_A, and type B subunits, N_B: $N = N_A + N_B$. The intramolecular composition is specified by the fractions $x_A = N_A/N$ and $x_B = N_B/N$. A random copolymer is characterized by a sequence distribution that follows Bernoullian statistics: the probabilities depend only on x_A and x_B.

As mentioned above, bulk random copolymers are often amorphous. If it is desired to increase the dielectric constant of a polymeric material, a copolymer that incorporates polar subunits in a random way will achieve this goal. Because the system now contains more than one kind of subunit, there will be fluctuations in the local composition, but these variations are expected to be small in magnitude and in spatial extent. A homopolymer that is glassy at room temperature can be converted to a copolymer with a lower glass-transition temperature, or exactly the reverse can be achieved. A crystalline homopolymer can be converted to an amorphous copolymer, or the melting temperature can be lowered by the introduction of intramolecular "solute." The range of modifications that can be introduced by creating a random copolymer is limited only by the imagination of the chemist.

When the sequence distribution is not random, new material structures can be observed. One class of such materials is based on so-called segmented copolymers. The sequence distribution is characterized by strong correlations in the local composition: there are long runs of A-type and B-type mers. The fluctuations in the composition of bulk samples are now greater in magnitude and in spatial extent. The presence of structural heterogeneity on the length scales observed in segmented copolymers can have beneficial effects on the mechanical properties of such systems. A special case of this type of macro-molecule is the block copolymer: the sequence distribution is characterized by blocks of a well-defined length. Consider a diblock copolymer: all the A subunits are at one end and all the B subunits occur at the other end. Bulk materials display a remarkable structure: **the composition displays regions of nearly pure A and pure B that are macroscopic in extent!**

The size and shape of these regions depends on the value of N, the value of x_A, and the temperature and pressure. At least one dimension of these regions remains microscopic and is comparable to the radius of gyration, R_G, of chains of pure A or B with Nx_A or Nx_B subunits. If the volume fractions of the two subunits are comparable, the material displays a lamellar structure with layers of A alternating with layers of B. The thickness of the layers remains molecular in extent, but the area can become macroscopic. If the volume fractions differ substantially, one phase can become continuous, with separated regions of the other phase embedded in the matrix. One of the most remarkable aspects of the structure of these systems is the regularity of the location of the separated phases. If the minority phase is dilute enough, the separated phases are spherical with radii that are molecular in extent. The spheres are found on a three-dimensional hexagonal lattice. At interme-diate volume fractions, the morphology of the separated phase is cylindrical, with the axes arranged on a two-dimensional hexagonal lattice.

The detailed morphology can be explained in terms of the creation of a minimum surface for the two-phase system: the Gibbs energy is minimized when the surface-to-volume ratio is a minimum. The phase regularity is reinforced by the molecular regularity of the diblock copolymer. In one remarkable range of composition, the morphology is bicontinuous, with both phases characterized by local dimensions that are molecular in extent. An electron micrograph of such a structure is shown in Figure 1.2. The field of block copolymers has continued to yield new materials for many years and shows great promise for the future.

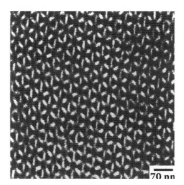

Figure 1.2 Bright-field transmission electron micrograph of a polystyrene-polyiso-prene block copolymer in the bicontinuous phase. (From Thomas, E.L. and Lescanec, R.L., Phase morphology in block copolymer systems, *Phil. Trans. R. London, A*, 348, 149–166, 1994. With permission.)

1.6   Suggestions for further reading

Even though it has now been in print for over 50 years, the essential single book for every physical chemist who desires to understand macromolecules is Flory's *Principles of Polymer Chemistry*.[15] There is a good introduction and considerable historical material in that volume.

An excellent current history of polymer science is *Polymers: the Origins and Growth of a Science*.[30] Many of the historical details and personal insights presented in this chapter were gleaned from that book.

chapter two

Describing polymer structure

2.1 Introduction

The first task in the description of a macromolecule is to enumerate all the atoms. The results can be summarized as a set: $\{A_i\}$. The molecular weight M for the polymer can be then be calculated as:

$$M = \sum_{i=1}^{N} M_i \qquad (2.1)$$

where the sum is over all N atoms in the molecule and M_i is the atomic weight for each atom. Although there are many macromolecules, especially biopolymers, for which the value of N is unique, many other macroscopic samples of polymeric material are characterized by a distribution of N and hence of M. The concept of a molecular weight distribution, $\rho(M)$, is central to the modern practice of polymer science. The consequences of such a distribution will be considered throughout this book.

Analysis of the properties of macromolecules is usually facilitated by identifying appropriate subunits and end groups. The molecular weight can then be expressed as:

$$M = \sum_{j=1}^{n} M_j + \sum_{k=1}^{e} M_k \qquad (2.2)$$

where n is the number of mers with molecular weight M_j, and e is the number of end groups with molecular weight M_k. Another term for n is the degree of polymerization. If $e = 0$, the structure is closed. If $e = 2$, the molecule is linear. And if $e > 2$, the molecule is branched. A distribution of the number of mers can also be defined: $\rho(n)$.

NEUROTENSIN

PYG LEU TYR GLU ASN LYS PRO ARG ARG PRO TYR ILE LEU

Figure 2.1 The amino acid sequence of a hypothalamic peptide neurotensin. The three-letter codes stand for the following amino acids: ARG (arginine), ASN (asparagine), GLU (glutamic acid), ILE (isoleucine), LEU (leucine), LYS (lysine), PRO (proline), PYG (pyroglutamic acid), and TYR (tyrosine). (Data from Carraway, R. and Leeman, S.E., *J. Biol. Chem.*, 250, 1907, 1975. With permission.)

If the macromolecule is a copolymer, the sequence of mers identified in the previous paragraph can be specified. Most proteins are characterized by a unique sequence of peptide subunits. The peptide sequence for a typical small protein is shown in Figure 2.1

As noted in Chapter 1, many synthetic copolymers have a distribution of mer sequences and mer compositions, as well as a distribution of molecular weight. The precise description of such a system requires many measures of the mer distribution. Detailed examples of copolymer composition and sequence will be considered throughout this text.

2.2   Geometric structure of macromolecules

The geometric structure of the molecule is defined by the locations of each of the atoms and by the specification of the chemical bonds. The atomic locations can be represented as a set of position vectors, $\{\vec{r}_i\}$. It is convenient to encode the chemical bond information as a set of bond vectors, $\{\vec{l}_j\}$, where each bond vector is calculated as the vector difference between the two position vectors for the bonded atoms: $\vec{l}_j = \vec{r}_{j+1} - \vec{r}_j$.

For a macromolecule, the amount of information contained in the set of bond vectors is large. One way to reduce the amount of information that must be considered in detail is to construct a subset of the bond vectors associated with a molecular backbone for the polymer. The set of bond vectors that connects each subunit to its neighbors and the bond vectors within each mer that connect the atoms involved in the chain structure are chosen as the backbone bonds. Other bond vectors between multivalent atoms within each mer are designated side-chain bonds. Consideration of the set of backbone bond vectors greatly reduces the conceptual task. The length of a backbone bond vector, l_j, does change as a function of time, but the variance of the bond length is a small fraction of its magnitude, and it is convenient to treat the backbone bond lengths as fixed quantities. It is also conventional to define bond angles, $\{\theta_j\}$, between the backbone bonds. The bond angle is chosen so that the scalar product of the bond vectors of bonds j and $j + 1$ is given by:

$$\vec{l}_j \cdot \vec{l}_{j+1} = l_j l_{j+1} \cos \theta_j .$$

(2.3)

BACKBONE FRAME OF REFERENCE

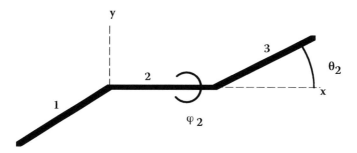

Figure 2.2 Discussion of the geometry of the chain backbone is facilitated by adopting a common frame of reference. Each pair of bonds is characterized by a bond angle θ_i. Each three-bond sequence is characterized by a bond rotation (dihedral) angle ϕ_i. The backbone bond vectors can be expressed in the common frame of reference by using transformation relations like that of Equation 2.4.

The values of the bond angles also change with time in a real molecule, but the variations are centered around a well-defined average value. In the following discussion, the bond angles will thus be treated as fixed.

The backbone conformation can be specified by a set of dihedral angles, $\{\phi_j\}$, between bonds $j - 1$ and $j + 1$. Consider the linear chain molecule backbone illustrated in Figure 2.2. Choose the location of the second backbone atom as the origin of coordinates and let the x-axis be defined by the direction of bond 2. The x–y-plane is defined by vectors $\vec{l_1}$ and $\vec{l_2}$. The z-axis is constructed to give a proper Cartesian coordinate system. The angle ϕ_2 is defined as the angle of rotation of $\vec{l_3}$ about the x-axis necessary to cause it to point in the x–y-plane in a direction comparable to $\vec{l_1}$. The angle ϕ_2 can also be specified in terms of the transformation of the vector $\vec{l_3}$ into the frame of reference of vector $\vec{l_2}$. In its own frame of reference, a vector is characterized by one variable, its length l_i. The components of $\vec{l_3}$ in our chosen frame of reference are:

$$l_{3x} = l_3 \cos \theta_2$$

$$l_{3y} = l_3 \sin \theta_2 \cos \phi_2 \qquad (2.4)$$

$$l_{3z} = l_3 \sin \theta_2 \sin \phi_2$$

2.3   Bond probability distributions

In a sample containing many macromolecules, each one can have a different set of dihedral angles for the backbone bonds. The internal energy of the chain molecule, $E(\{\phi_j\})$, depends on the conformation described by the backbone rotation angles. The probability of any particular set depends on the energy of the polymer in that conformation. The probability can be expressed as:

$$\wp\left(\{\phi_j\}\right) = \exp\left[-E\left(\{\phi_j\}\right)/k_bT\right]/Z$$

$$Z = \int\limits_{\text{all}\{\phi_j\}} \exp\left[-E\left(\{\phi_j\}\right)/k_bT\right]d\{\phi_j\}$$

(2.5)

where Z is called the conformation partition function and the integral is over all sets of dihedral angles. While Equation 2.5 is an exact formulation, within the context of the backbone bond description, it is not useful for carrying out actual calculations of polymer properties. The calculation of the full energy function for a typical macromolecule is a daunting task. The evaluation of the multidimensional integral is even more problematic. Many simplifying approaches have been employed to treat the statistics of chain molecules. This chapter presents three of the most common schemes:

1. Rotational isomeric state (RIS) approximation
2. Random coil model
3. Persistent (wormlike) chain model

2.4   *Rotational isomeric state approximation*

The energy of a flexible molecule is a function of the dihedral angles between pairs of bond vectors separated by the bond about which rotation takes place. In discussing this issue, consider the molecule *n*-butane. There are four backbone atoms and three bond vectors to consider. Rotation about bonds 1 and 3 produces no change in the conformation of the chain backbone. Only changes in ϕ_2 lead to a different conformation. The energy of the molecule as a function of ϕ_2 is plotted in Figure 2.3.

The energy is at a minimum when $\phi_2 = 0$ and all three bond vectors lie in the same plane. This value of ϕ_2 is called the *trans* state. The energy is at an absolute maximum when $\phi_2 = \pi$. This conformation is called the *cis* state. Local minima are observed when the dihedral angle is near $\pm(2\pi/3)$. These conformations are called *gauche* states. An empirical function that represents the curve in Figure 2.3 is:

$$E\left(\phi\right) = A + B\left(1 - \cos 3\phi\right) + C\sin\left(\phi/2\right)$$

(2.6)

The probability of observing a butane molecule with a particular conformation is given by Equation 2.5 for the case of a molecule with a single internal rotation angle.

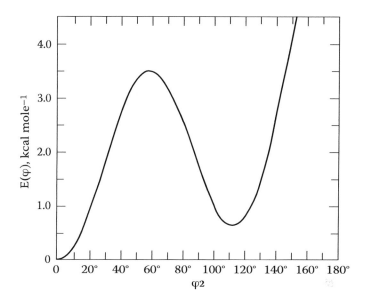

Figure 2.3 The conformational energy of *n*-butane is a function of the internal rotation angle ϕ_2. (From Flory, P.J., *Statistical Mechanics of Chain Molecules*, Interscience, New York, 1969. With permission.)

Many of the properties of the molecule depend on its conformation. Consider some typical property: $A\left(\left\{\vec{l}_j\right\}\right)$. The average value of this property for an equilibrium ensemble of molecules is

$$\langle A \rangle = \int_{\text{all}\left\{\vec{l}_j\right\}} A\left(\left\{\vec{l}_j\right\}\right) \wp\left(\left\{\vec{l}_j\right\}\right) d\left\{\vec{l}_j\right\} \tag{2.7}$$

where $\wp\left(\left\{\vec{l}_j\right\}\right)$ is the probability for a particular conformation and the integral is over all conformations. While such an expression can be evaluated explicitly for simple molecules, difficulties are encountered for macromolecules. The rotational isomeric state approximation consists in replacing the integral over a continuous set of conformations with a sum over a discrete set of conformations. Suppose we divide the continuous range of some dihedral angle, ϕ_j, into three parts:

1. Centered about the *trans* state and including the value of the first local maximum, designated as *t*
2. From the end of the *t* range to the *cis* state for larger values of ϕ_j, designated as g^+
3. The comparable range for smaller values of ϕ_j, designated as g^-

These ranges are called rotational isomeric states (RIS). Any conformation of the molecule can be assigned a particular set of RIS values, $\{s_j\}$. The average can then be expressed as:

$$\langle A \rangle = \sum_{\text{all}\{s_j\}} A\left(\{s_j\}\right) \wp\left(\{s_j\}\right) \tag{2.8}$$

where the sum is over all sets of rotational isomeric states. The obvious ambiguity is the calculation of the value of the property, A, in any particular RIS state, since there will still be a large number of conformations in that state. A direct calculation of the average value of A in the RIS state is just as difficult as the original calculation that we are trying to avoid. The conventional approximation used in the RIS scheme is to use the value of A for a particular value of ϕ in each rotational isomeric state. For the t state, the value $\phi_t = 0$ is usually chosen. For the two g states, the value of ϕ_g is often treated as an adjustable parameter, but it must be near $2\pi/3$.

Once the rotational isomeric states have been chosen, it is necessary to assign probabilities to each set of $\{s_j\}$. Consider the molecule n-butane. The RIS probabilities for the g states must be equal. If the relative probability of a g state is denoted as σ, the absolute probabilities are:

$$\wp(t) = 1/\left(1 + 2\sigma\right)$$
$$\wp(g^{\pm}) = \sigma/\left(1 + 2\sigma\right) \tag{2.9}$$

Calculation of the RIS probabilities for molecules with more than one internal rotation introduces the concept of neighbor-dependent conformational energies. Consider the molecule n-pentane. There are now four backbone bonds. The energy of the molecule will depend on the values of ϕ_2 and ϕ_3 taken together. The total conformational space of the molecule can be divided into nine rotational isomeric states, based on the n-butane states. The nine states will be denoted: tt, tg^+, tg^-, g^+t, g^-t, g^+g^+, g^-g^-, g^+g^-, and g^-g^+. The relative probability of any of the four states involving only one t will be denoted σ, with the tt state taken as 1. It is convenient to express the relative probability for the two states with identical g rotations as $\sigma^2\psi$, where the value of ψ is chosen to express the exact result. The value of ψ can be either greater than 1 or less than 1, depending on the polymer being considered. If $\psi = 1$, then the two rotations can occur in the same direction without any additional energy associated with the occurrence of that specific sequence. When the states $g^{\pm}g^{\mp}$ are considered, it is clear from molecular models that the end groups are brought close to each other. The relative probability of these states is written as $\sigma^2\omega$. The value of ω is usually very small. The absolute probabilities for the nine states are:

$$\wp\left(tt\right)=1/\left(1+4\sigma+2\sigma^2\psi+2\sigma^2\omega\right)=1/Z$$

$$\wp\left(tg^{\pm}\right)=\wp\left(g^{\pm}t\right)=\sigma/Z$$

$$\wp\left(g^{\pm}g^{\pm}\right)=\sigma^2\psi/Z \tag{2.10}$$

$$\wp\left(g^{\pm}g^{\mp}\right)=\sigma^2\omega/Z$$

While it might be supposed that the statistics of *n*-hexane would be even more complicated, it is found empirically that consideration of bond pairs is sufficient to account for the observed probabilities. For example, the relative probability of the *tg⁺t* state is well described by the same σ used above. Because the approximation of nearest-neighbor interaction energies is sufficient to calculate the probabilities, sophisticated mathematical methods can be used in the evaluation of average properties. The interested reader is referred to *Statistical Mechanics of Chain Molecules*.[19]

2.5 Mean-squared end-to-end distance

To illustrate the calculation of a geometric average for a chain molecule, the mean-squared end-to-end distance $\langle R^2 \rangle$ will be considered. The end-to-end vector $\vec{R}$ is defined as the sum of all the backbone bond vectors for a linear chain:

$$\vec{R}=\sum_{j=1}^{n}\vec{l}_j . \tag{2.11}$$

Each rotational isomeric state for a chain molecule can be assigned a particular value of $\vec{R}$. The mean-squared end-to-end distance is defined as:

$$\langle R^2 \rangle = \langle \vec{R}\cdot\vec{R} \rangle = \sum_{j,k=1}^{n}\langle \vec{l}_j \cdot \vec{l}_k \rangle \tag{2.12}$$

where the brackets denote an average over all RIS states and the double sum is over all pairs of bond vectors. To execute the scalar product between the two bond vectors, they must be expressed in the same coordinate system. The formulation of this procedure and the calculation of the averages is described in detail in the previously noted book by Flory.[19]

To illustrate the calculation of conformational averages in the rotational isomeric state approximation, the molecules *n*-butane and *n*-pentane will be considered. The end-to-end vector for the t state of n-butane in the frame of reference developed above can be expressed as:

$$\vec{R}_t = l_{cc} \begin{bmatrix} 1+2\cos\theta \\ 2\sin\theta \\ 0 \end{bmatrix} \tag{2.13}$$

where l_{cc} is the length of the C-C bond. The corresponding quantity for one of the g states is:

$$\vec{R}_g = l_{cc} \begin{bmatrix} 1+2\cos\theta \\ \sin\theta(1+\cos\phi) \\ \sin\theta\sin\phi \end{bmatrix}. \tag{2.14}$$

The corresponding mean-squared quantities can be evaluated if particular bond angles and rotation angles are selected. To facilitate the calculations, the bond angle will be assigned its tetrahedral value, which is approximately 70.5°. The values of the trigonometric functions are then:

$$\cos\theta = 1/3 \text{ and } \sin\theta = 2\sqrt{2}/3 .$$

The dihedral angle for the g^+ state will be taken as exactly $(2\pi/3)$. The corresponding quantities are:

$$\cos\phi = -1/2 \text{ and } \sin\phi = \sqrt{3}/2 .$$

The squared end-to-end distances for the two states are:

$$R_t^2 = l_{cc}^2\left(19/3\right) \text{ and } R_g^2 = l_{cc}^2\left(33/9\right). \tag{2.15}$$

The mean-squared end-to-end distance is:

$$\left\langle R^2 \right\rangle = R_t^2 \wp_t + 2R_g^2 \wp_g . \tag{2.16}$$

The end-to-end vectors for the nine states of n-pentane can be calculated by adding the appropriate vector to the expression for $\vec{R}$ derived for n-butane. Consider the tt state. The fourth bond vector points in the same direction as the second bond vector. In the frame of reference of the second bond vector, the end-to-end vector for the tt state is:

$$\vec{R}_{tt} = l_{cc} \begin{bmatrix} 2 + 2\cos\theta \\ 2\sin\theta \\ 0 \end{bmatrix}. \tag{2.17}$$

The length of the end-to-end vector for the four states containing one g state is the same. It is most convenient to calculate the end-to-end vector for the $g^{+}t$ state by considering a molecular model. When the third bond is in the *trans* conformation, the direction of the second and fourth bond vectors is the same. The end-to-end vector can then be calculated by adding the fourth bond vector to the expression derived for the g^{+} state of *n*-butane:

$$\vec{R}_{g^{+}t} = l_{cc} \begin{bmatrix} 2 + 2\cos\theta \\ \sin\theta(1+\cos\phi) \\ \sin\theta\sin\phi \end{bmatrix}. \tag{2.18}$$

The length of the end-to-end vector for the states with identical g states is also the same. The fourth backbone bond vector for the $g^{+}g^{+}$ state points in the opposite direction from the third backbone bond vector in the g^{-} state of *n*-butane. For a definition of the rotation angle, which is still positive, the fourth bond vector can be expressed as:

$$\vec{l}_{4} = -l_{cc} \begin{bmatrix} \cos\theta \\ \sin\theta\cos\phi \\ -\sin\theta\sin\phi \end{bmatrix}. \tag{2.19}$$

The end-to-end vector for the $g^{+}g^{+}$ state can then be calculated by adding the expression for $\vec{l}_{4}$ to the end-to-end vector for the g^{+} state of *n*-butane:

$$\vec{R}_{g^{+}g^{+}} = l_{cc} \begin{bmatrix} 1 + \cos\theta \\ \sin\theta \\ 2\sin\theta\sin\phi \end{bmatrix}. \tag{2.20}$$

The fourth bond vector for the $g^{+}g^{-}$ state points in the opposite direction of the first bond vector. In the frame of reference of the second bond vector, the end-to-end vector for this state is:

$$\vec{R}_{g^{+}g^{-}} = l_{cc} \begin{bmatrix} 1 + \cos\theta \\ \sin\theta\cos\phi \\ \sin\theta\sin\phi \end{bmatrix}. \tag{2.21}$$

The four values of $R_{ss'}^2$ are then:

$$R_{tt}^2 = l_{cc}^2 \left(32 / 3 \right)$$

$$R_{tg^{\pm}}^2 = l_{cc}^2 \left(8 \right)$$

$$R_{g^{\pm}g^{\pm}}^2 = l_{cc}^2 \left(16 / 3 \right) \tag{2.22}$$

$$R_{g^{\pm}g^{\mp}}^2 = l_{cc}^2 \left(8 / 3 \right)$$

The mean-squared end-to-end distance for n-pentane can be expressed as:

$$\left\langle R^2 \right\rangle = R_{tt}^2 \wp_{tt} + 4 R_{tg}^2 \wp_{tg} + 2 R_{g^{+}g^{+}}^2 \wp_{g^{+}g^{+}} + 2 R_{g^{+}g^{\mp}}^2 \wp_{g^{+}g^{\mp}} . \tag{2.23}$$

The reader should verify the results for n-pentane.

The evaluation of $\left\langle R^2 \right\rangle$ has been carried out for all the n-alkanes from $n = 4$ to $n = 250$. It is convenient to define the characteristic ratio, C_n, for a chain molecule as:

$$C_n = \frac{\left\langle R^2 \right\rangle}{n \left\langle l^2 \right\rangle} \tag{2.24}$$

where $\left\langle l^2 \right\rangle$ is the mean-squared length of a backbone bond. The value of C_n rises for small values of n, but it reaches a constant value at large ($n > 250$) degrees of polymerization. The asymptotic linear dependence of $\left\langle R^2 \right\rangle$ on the chain length has profound implications. When the chain is long enough, the global geometric properties can be described by a purely statistical model. A chain molecule that has reached the asymptotic limit is called a **random coil macromolecule.**

2.6   Statistics of the random coil

A transparent model can be derived from the backbone bond description given above to illustrate the properties of the random coil. The backbone of n bonds is divided into m equal groups, where n/m exceeds the degree of polymerization required to achieve the asymptotic characteristic ratio. Each subgroup is characterized by its own end-to-end vector, $\vec{r}_k$, where:

$$\vec{r}_k = \sum_{j=1+(k-1)n/m}^{kn/m} \vec{l}_j . \tag{2.25}$$

The end-to-end vector for the entire chain can then be expressed as:

$$\vec{R} = \sum_{k=1}^{m} \vec{r}_k .$$

(2.26)

where the sum is over all m subgroups. The mean-squared end-to-end distance is given by:

$$\langle R^2 \rangle = \sum_{k,l=1}^{m} \langle \vec{r}_k \cdot \vec{r}_l \rangle = m \langle r^2 \rangle$$

(2.27)

where $\langle r^2 \rangle$ is the mean-squared end-to-end distance of the backbone bonds in the subgroup. Since each subgroup contains many backbone bonds, the directions of the end-to-end vectors for any pair of subgroups are uncorrelated, and the average over all conformations of the scalar product is:

$$\langle \vec{r}_k \cdot \vec{r}_l \rangle = \langle r^2 \rangle \delta_{kl}$$

(2.28)

where δ_{kl} is the Kroneker delta, which equals 1 if $k = l$ and equals 0 if $k \neq l$. The loss of correlation in the direction of backbone bonds is the signature of the random coil.

Another important measure of the statistics of the random coil is the probability distribution for the end-to-end vector, $\wp(\vec{R})$. In the long-chain limit, this function is:

$$\wp(\vec{R}) = \left(\frac{3}{2\pi \langle R^2 \rangle} \right)^{3/2} \exp\left(\frac{-3R^2}{2\langle R^2 \rangle} \right).$$

(2.29)

The distribution is Gaussian. The probability distribution for the end-to-end vector is normalized so that:

$$\int_{\text{all } \vec{R}} \wp(\vec{R}) d\vec{R} = 1$$

(2.30)

where the integral is over all end-to-end vectors. Since the Gaussian distribution is a function of R^2, it is convenient to express the vector differential as $d\vec{R} = 4\pi R^2 dR$. The integral can then be evaluated over all scalar magnitudes of the end-to-end vector. The probability distribution for the length of the end-to-end vector, $\wp(R)$, is:

$$\wp(R) = 4\pi R^2 \left(\frac{3}{2\pi \langle R^2 \rangle}\right)^{3/2} \exp\left(\frac{-3R^2}{2\langle R^2 \rangle}\right). \tag{2.31}$$

A macromolecule that can be described by these statistics is called a **Gaussian chain**.

The existence of an analytic form for the probability distribution means that any function of the end-to-end vector can be averaged over the conformations of a random coil chain:

$$\langle F(\vec{R}) \rangle = \int_{\text{all } R} F(\vec{R}) \wp(\vec{R}) d\vec{R}. \tag{2.32}$$

It can be easily verified that the mean-squared end-to-end distance can be obtained by applying Equation 2.32 to the function R^2. Another important quantity is the most probable value for the length of the end-to-end vector. It can be obtained by solving for the maximum value of $\wp(R)$ and is equal to:

$$R_{mp} = \left(\frac{2\langle R^2 \rangle}{3}\right)^{1/2}. \tag{2.33}$$

The average length of the end-to-end vector $\langle R \rangle$ is obtained from:

$$\langle R \rangle = \int_0^\infty 4\pi R^3 \wp(\vec{R}) dR = \left(\frac{8\langle R^2 \rangle}{3\pi}\right)^{1/2}. \tag{2.34}$$

Frictional properties of the polymer often depend on the quantity $\langle 1/R \rangle$, which is calculated as:

$$\langle 1/R \rangle = \int_0^\infty 4\pi R \wp(\vec{R}) dR = \left(\frac{6}{\pi \langle R^2 \rangle}\right)^{1/2}. \tag{2.35}$$

The reader should verify these results with the help of a table of Gaussian integrals.

2.7   Chain flexibility and the persistence length

The chain-length dependence of the characteristic ratio, when the degree of polymerization is below the asymptotic limit, can also be modeled with fewer parameters than are needed to implement the rotational isomeric state scheme. The local details of the chain structure can be characterized by a unique parameter. The persistence length a can be calculated from the back-bone bond vectors as:

$$a = \lim_{n \to \infty} \left\langle \vec{l_1} \cdot \vec{R} \right\rangle / l_1 = \left(1 / l_1\right) \sum_{j=1}^{\infty} \left\langle \vec{l_1} \cdot \vec{l_j} \right\rangle . \tag{2.36}$$

Eventually, there is no correlation in direction between the first bond vector and the jth bond vector, and the sum reaches its asymptotic value.

The other dimensional measure of the chain molecule is its contour length L. For a chain of intermediate contour length, $a < L < 50a$, it can be shown that the mean-squared end-to-end distance is given by:

$$\left\langle R^2 \right\rangle = 2La - 2a^2 \left[1 - \exp\left(-L / a\right) \right] . \tag{2.37}$$

Polymer molecules in this regime are called semiflexible or wormlike chains.

chapter three

Measuring polymer structure

3.1 Introduction

Polymer structure can be described at a series of hierarchical levels:

1. The primary structure includes the chemical connectivity of the macromolecule and is often an invariant quantity.
2. The secondary structure includes the local geometric parameters that define the three-dimensional conformation of the chain.
3. The tertiary structure includes the long-range interactions that define the global shape of the macromolecule.

A complete presentation of all the methods used to determine polymer structure at all these levels is best reserved for technical monographs. In the present text, three experimental techniques have been selected to illustrate the issues involved in measuring polymer structure. The measurement of primary structure is demonstrated using nuclear magnetic resonance (NMR) spectroscopy. The local conformation of a chain molecule is related to vibrational spectroscopy. And the global conformation of a chain molecule is derived from measurements of the scattering function, $S(q)$.

3.2 Polymer composition and sequence distribution

The peptide sequence in a protein can now be determined with automated equipment. A standard reference on this technique is the *Handbook of Protein Sequence Analysis*.[8] However, many synthetic copolymers cannot be taken apart with such precision; they must be analyzed as intact chains. Consider an AB copolymer. The composition can be specified by the mole fraction $x_A = 1 - x_B$. While some synthetic copolymers have an identical sequence distribution for each chain, most chains are characterized by sequence probability distributions, $\wp\left(\left\{A_i B_j\right\}\right)$. One example of a unique sequence distribution is the nearly perfectly alternating copolymer created by the polymerization of styrene and maleic anhydride. Block copolymers are another clear example of a nearly perfect sequence distribution.

Monosubstituted vinyl polymers, $-(CH_2CHX)_n-$, are actually stereochemical copolymers. The stereochemical configurations of the substituted carbon atoms can be considered in pairs. When the chain is viewed in the all-*trans* conformation, the side chains appear either in front of the main chain or behind the chain backbone. When the side chains on two successive asymmetric carbons appear on the same side of the chain backbone, the configuration is called **meso** (*m*). When the side chains occur on opposite sides of the chain backbone, the configuration is called **racemic** (*r*). The stereochemical configuration for a particular chain can be specified by the dyad sequence: e.g., *mmrmr*, ..., *rmrmmr*. If a chain is described by a stereochemical sequence consisting of all meso dyads, it is called isotactic. The corresponding sequence with all racemic dyads is called syndiotactic. Atactic polymers are random copolymers characterized by equal composition of *m* and *r* dyads. Because the properties of vinyl polymers can depend dramatically on the stereotactic composition and sequence, measurement of these quantities is essential for a complete understanding.

Polypropylene has been studied in great detail, and the stereotactic composition and sequence have been measured by NMR spectroscopy. The following section presents the essential principles of NMR spectroscopy needed to understand the interpretation of the spectra of polypropylene. The technical details are based on the book *Chain Structure and Conformation of Macromolecules*.[4]

3.3   *NMR spectroscopy of polymers*

One of the fundamental properties of atomic nuclei is nuclear spin. Each nucleus is characterized by a spin quantum number, I, and a magnetogyric ratio, γ. There are $2I + 1$ nuclear spin states for each nucleus with energies given by:

$$\varepsilon_m\left(H\right) = -m\gamma H \tag{3.1}$$

where m is the magnetic quantum number, which runs from $-I$ to I in integral steps, and H is the local magnetic field at the nucleus of the atom. The local magnetic field depends on the applied magnetic field and the chemical environment of the atom. It is this exquisite sensitivity to the precise local details of the electron density that makes NMR spectroscopy such a powerful tool in the analysis of the chemical structure of molecules. The potential power of NMR spectroscopy has been realized experimentally. Modern spectrometers combine high sensitivity to rare species with high resolution of different chemical environments. For example, end-group analysis is now the method of choice for determining the molecular weight of soluble macromolecules in the low to moderate range, since NMR can detect the rare end groups.

One of the most useful nuclei to study in the analysis of hydrocarbon structure is ^{13}C. It is characterized by $I = 1/2$. Experimental techniques have been developed that allow isolated ^{13}C nuclei to be observed, unperturbed by interactions with the ubiquitous ^{1}H nuclei. The energy difference between the $m = 1/2$ and $m = -1/2$ state is measured for all the ^{13}C nuclei in the sample. There are three types of carbon atom locations in the polypropylene molecule:

1. Methyl (CH_3) carbons in the side chain
2. Methylene (CH_2) carbons in the main chain
3. Methine (CH) carbons

Within each group, the exact local magnetic field experienced by each nucleus depends on the stereochemical sequence. The ^{13}C spectrum of a nearly perfect isotactic polypropylene is shown in Figure 3.1. There are only three peaks corresponding to the three kinds of carbon locations for chains characterized by pentad sequences that are *mmmm*. It is much more difficult to obtain nearly perfect syndiotactic polypropylene, but the spectrum of highly syndiotactic material is shown in Figure 3.1. The spectrum again separates into three regions, with one dominant peak in each corresponding to the *rrrr* pentad sequence. The spectrum of atactic polypropylene is considerably more complicated. An expanded presentation of the spectrum corresponding to the methyl carbons is shown in Figure 3.2. There are nine distinguishable peaks that have been assigned to the pentad sequences: *mmmm, mmmr, rmmr, rrmm, mrmm, rmrm, mrrm, rrrm,* and *rrrr*. The pentad assignments were confirmed by studying carefully synthesized oligomers. The close collaboration between synthetic chemists and physical chemists on this problem is one of the triumphs of polymer science.

High-resolution NMR is usually carried out on polymer solutions of moderate concentration with macromolecules of low to moderate molecular weight. It is important that the molecules reorient fast enough to produce an average environment at each nucleus. Solid samples have been studied by mechanically rotating the NMR probe at high speed oriented at the magic angle (the angle at which $3 \cos^2\theta = 1$). Polymer chains in solution also undergo local conformational isomerizations at a rate that usually exceeds the NMR frequency for ^{13}C spectroscopy. Only the fixed chemical structure does not average out under these conditions. NMR spectroscopy is the method of choice for studying the composition and sequence distribution for macromolecules.

3.4 Measuring local conformations of polymers

The conformational state of a chain molecule can be specified by a set of bond lengths, $\{l_j\}$, a set of bond angles, $\{\theta_j\}$, and a set of rotation angles for

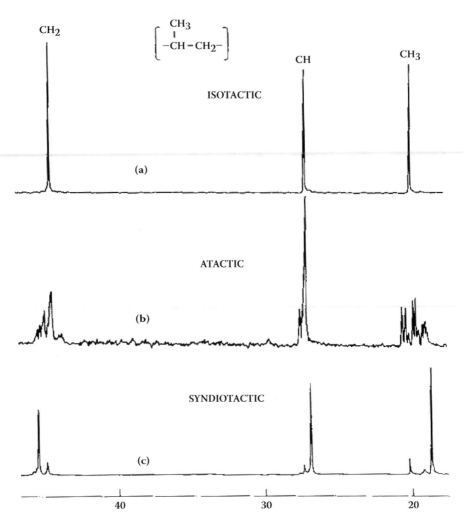

Figure 3.1 The 25-MHz ^{13}C spectra of three preparations of polypropylene: isotactic, atactic, and syndiotactic. (From Bovey, F.A., *Chain Structure and Conformation of Macromolecules*, Academic Press, New York, 1982. With permission.)

the single bonds of the chain backbone, $\{\phi_k\}$. In the liquid state, the conformation of any particular molecule is constantly changing. However, there will be a probability distribution for the conformational state of every bond, and this can be used to define the average properties of the macromolecular system. Consider the homologous series of *n*-alkanes described in Chapter 2. Each molecule was assigned to a particular rotational isomeric state. For example, the molecule *n*-hexane can be described by 27 rotational isomeric states (RIS). However, is there any way to observe these different groups of molecular conformations in the laboratory?

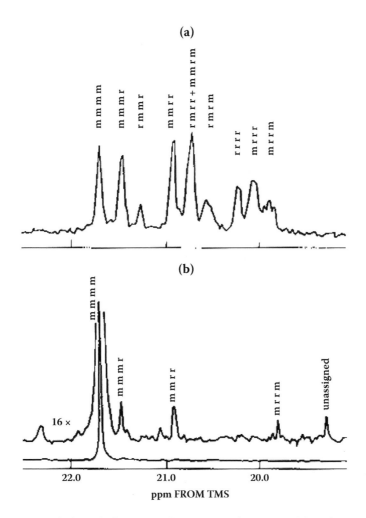

Figure 3.2 Expanded methyl region of Figure 3.1 for atactic (a) and isotactic (b) polypropylene. (From Bovey, F.A., *Chain Structure and Conformation of Macromolecules*, Academic Press, New York, 1982. With permission.)

To observe particular rotational isomeric states, the method must be much more rapid than the rate of conformational isomerization. Optical methods such as absorption spectroscopy or light-scattering spectroscopy provide a short-time probe of the molecular conformation. If the electronic states of the molecule are strongly coupled to the backbone conformation, the ultraviolet or visible spectrum of the molecule can be used to study the conformational composition. The vibrational states of macromolecules are often coupled to the backbone conformation. The frequencies of molecular vibrations can be determined by infrared absorption spectroscopy and Raman scattering spectroscopy. The basic principles of vibrational spectros-

copy are presented and applied to the study of the conformational states of
n-hexane in Section 3.5.

3.5 *Vibrational spectroscopy*

The essential signature of a molecule is that it vibrates. For a molecule
composed of N atoms, there are $3N$ mechanical degrees of freedom associated
with the motions of the system. Three degrees of freedom are determined
by the translational motions of the center of mass, and for a nonlinear
molecule there are three degrees of freedom connected with the overall
rotational motion of the molecule. For a macromolecule, some of the remain-
ing $3N - 6$ degrees of freedom are associated with isomerizations of the chain
backbone and the side chains. Finally, there exists a set of quantized vibrational states for the molecule. If the frequencies of the vibrational states
depend on the conformational state of the molecule, the measurement of the
vibrational spectrum can be used to infer the conformational composition
of the ensemble of macromolecules. The frequencies of quantized molecular
vibrations greatly exceed the frequencies associated with isomerization of
the chain backbone.

Molecular vibrational motion can be observed in two ways. A molecule
is characterized by an electric dipole moment, $\bar{\mu}$. As the system vibrates, the
components of this vector change with time. If dipole radiation of the same
frequency is present, the molecular system can exchange energy with the
optical field. If the vibrational modes are in the ground state, the molecule
absorbs infrared light with the same frequencies as those that characterize the
vibrational motions of the system. A molecule is also characterized by an
optical polarizability tensor, $\bar{\alpha}$. As the molecule vibrates, the components of
the tensor change with time. Light incident on this system can be scattered
by the molecule and can exchange energy to give scattered light with frequencies that differ from the incident frequency by the characteristic frequency of
the molecular vibration. This process is called Raman scattering. The two
techniques are complementary, since the modulation of the molecular dipole
moment and the molecular polarizability depend on different measures of the
electronic structure of the molecule. Some vibrations may be weakly visible
by infrared (IR) spectroscopy but very detectable by Raman spectroscopy.

The experimental aspects of IR spectroscopy have reached a very high
level of sensitivity and resolution. The transmission of infrared light of
well-defined frequency is measured along a reference path and along the
sample path. The relative transmission is determined as a function of IR
frequency. In a Raman scattering experiment, a laser beam of high intensity
and well-defined frequency is incident on the sample. In general, no reference beam is employed. The scattered light is collected over a large solid
angle and analyzed with a double monochromator. The light intensity is
measured as a function of absolute frequency. The absolute frequency is then
converted to the frequency shift by subtracting the absolute frequency of the
incident light. The measured intensity is then plotted against the frequency

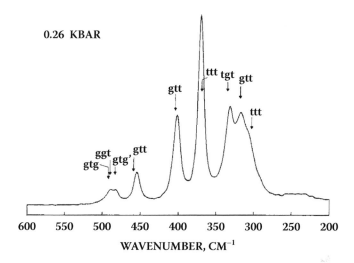

0.26 KBAR

Figure 3.3 Raman spectrum of *n*-hexane showing the bands due to distinct conformers. (From Snyder, R.G., Infrared and raman spectra of polymers, *J. Chem. Phys.*, 47, 1316, 1967. With permission.)

shift to yield the Raman spectrum. An excellent article on the application of IR and Raman spectroscopy to polymers is found in the literature.[38]

3.6   Conformational statistics of n-hexane

The vibrational modes of the molecule *n*-hexane have been analyzed in detail by Snyder[43] as a function of the conformation of the chain. Distinct sets of vibrational frequencies have been calculated for the ten distinguishable conformers of the molecule. In addition, the relative intensities of the Raman peaks have been calculated. The Raman spectrum of *n*-hexane in the region from 200 to 600 cm^{-1} is shown in Figure 3.3. Distinct peaks associated with the *ttt*, *gtt*, and *tgt* conformations were identified. Features associated with conformations containing more than one *g* state were less well resolved and much lower in intensity.

3.7   Global measures of the chain structure

The mean-squared end-to-end distance depends on an average over all the conformations of a chain molecule and gives a single-number measure of the global extent of the macromolecule. Another important quantity that depends on an average over the conformations of the whole chain is the mean-squared radius of gyration $\langle R_G^2 \rangle$. Each macromolecule is characterized by the location of its center of mass, $\vec{r}_{cm}$. In the frame of reference of the center of mass, the location of each atom can be expressed as a vector, $\vec{s}_i$, that is equal to the vector difference between the location of the atom and the center of mass:

$$\vec{s}_i = \vec{r}_i - \vec{r}_{cm} \, . \tag{3.2}$$

The mean-squared radius of gyration is the mass-weighted average squared distance of an atom in the macromolecule from its center of mass:

$$\left\langle R_G^2 \right\rangle = \left(1/M\right) \sum_{i=1}^{N} M_i \left\langle \vec{s}_i \cdot \vec{s}_i \right\rangle . \tag{3.3}$$

For a homopolymer, it is convenient to calculate the mean-squared radius of gyration by dividing the chain into the same m subunits discussed in Section 2.6 for the random coil chain. Because all of the subunits have the same weight, the mass weighting ceases to be necessary. There is also a theorem due to Lagrange that allows $\left\langle R_G^2 \right\rangle$ to be expressed in terms of the intersubunit vectors, $\vec{R}_{kl}$:

$$\left\langle R_G^2 \right\rangle = \left(1/2\right)\left(1/m^2\right) \sum_{k \neq l=1}^{m} \left\langle \vec{R}_{kl} \cdot \vec{R}_{kl} \right\rangle . \tag{3.4}$$

where the double sum is over all unequal pairs of subunits and the terms are given by the mean-squared distance between subunits k and l. The mean-squared radius of gyration for a homopolymer can be measured by light scattering. A detailed discussion of this technique can be found in the book *Light Scattering from Polymer Solutions.*[22]

3.8 Light scattering from dilute polymer solutions

The light-scattering experiment consists of shining a well-defined beam of light of wavelength λ on a sample of refractive index n_0. Scattered light is observed along a well-defined direction that makes an angle θ with the incident direction. The intensity of scattered light is measured as a function of the scattering angle $I(\theta)$. Light scattering is characterized by an inverse distance q given by:

$$q = \left(\frac{4\pi n_0}{\lambda}\right) \sin\left(\theta/2\right) . \tag{3.5}$$

The scattering process is also characterized by a scattering vector $\vec{q}$ whose length is equal to q and whose direction is given by the vector difference between a unit vector in the scattering direction and a unit vector in the incident direction.

Light is scattered by all fluids that do not fully absorb the incident light. To carry out measurements that can be properly interpreted, the fraction of the incident light that is scattered should be small. The intensity of scattered light from a dilute polymer solution exceeds the intensity for the pure solvent. Thus the difference in the polarizability density for the polymer solute and the solvent leads to optical inhomogeneities that scatter light. The excess scattered intensity is calculated for a constant volume of observation as a function of scattering angle. The polymer scattering function $S(q)$ is defined as:

$$S(q) = \left(\frac{\Delta I(\theta)\sin\theta}{\lim_{\theta \to 0} \Delta I(\theta)\sin\theta} \right)$$

(3.6)

where $\Delta I = I_{solution} - I_{solvent}$. The scattering function depends on a globally averaged measure of the chain conformation.

The theoretical description of the scattering function is facilitated by dividing the chain into m identical subunits characterized by an excess polarizability density. All the subunits are identical in composition, so it is not necessary to weight the result by the local polarizability density. The scattering function can then be expressed as:

$$S(q) = \left(1 / m^2\right) \sum_{k,l=1}^{m} \left\langle \exp\left(i\vec{q} \cdot \vec{R}_{kl}\right) \right\rangle$$

(3.7)

where the brackets denote an average over all conformations of the macro-molecule and over all orientations in the laboratory. If the probability distribution for the intersubunit vectors is given by the Gaussian function (Equation 2.29), the scattering function can be calculated analytically. An excellent discussion of this process is found in the book *Dynamic Light Scattering*.[3] The result can be expressed as:

$$S(q) = \left(2 / u^2\right)\left[u - 1 + \exp\left(-u\right) \right]$$

(3.8)

where $u = q^2 \left\langle R_G^2 \right\rangle$. A graph of $S(q)$ plotted against u is presented in Figure 3.4.

To appreciate the overall shape of $S(q)$, it is necessary to make measurements at large values of u. If the only quantity that is desired is the mean-squared radius of gyration, it is possible to consider data only in the range where $u < 1$. A mathematical expansion of the scattering function in the low-u regime yields:

$$S(q) = 1 - \left(q^2 / 3\right)\left\langle R_G^2 \right\rangle + \cdots .$$

(3.9)

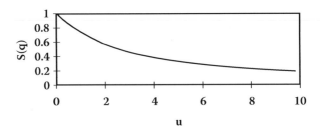

Figure 3.4 The scattering function for a Gaussian chain is called the Debye function. It is a universal function of the scaled variable u. The value of the scattering function for a single particle starts at 1 for $q = 0$ and decreases at higher scattering angles.

Measurements of the excess scattered intensity of a dilute polymer solution can then be used to calculate the mean-squared radius of gyration for the macromolecule. The result shown in Equation 3.9 is quite general for macromolecules that can be divided into identical subunits. If the macromolecule is a Gaussian chain, $\langle R_G^2 \rangle = (1/6) \langle R^2 \rangle$.

chapter four

The macromolecular basis of rubber elasticity

4.1 Introduction

The existence of the rubbery state of polymers was introduced in Chapter 1. The present chapter provides a more formal development of the properties of rubber and explains the observed phenomena in terms of the structural properties of macromolecules developed in Chapter 2. A thorough understanding of the behavior of rubber is essential to explain the properties of macromolecules in dilute solution, which is the subject of Chapter 5.

The development of the physical chemistry of rubber was greatly aided by the clear definition of an "ideal" state for this material. An ideal rubber is an amorphous, isotropic solid. The liquidlike structure of rubber was discovered very soon after the technique of X-ray scattering was developed. An isotropic material is characterized by physical properties that do not depend on the orientation of the sample. The deformation of an isotropic solid can be characterized by only two unique moduli: the modulus of compression, K, and the shear modulus, G. A solid is characterized by equilibrium dimensions that are functions of temperature, pressure, and the externally imposed constraints. It is convenient to define a shape vector, $\vec{L}$, whose components are the length, width, and height of a rectangular parallelepiped. For a system with no external constraints, the shape vector can be expressed as:

$$\vec{L}_0 = \begin{bmatrix} L_{x0} \\ L_{y0} \\ L_{z0} \end{bmatrix}.$$ (4.1)

The volume of the sample is simply $V = L_x L_y L_z$. The modulus of compression for a sample of rubber is comparable to that for a liquid. The equilibrium shear modulus for a liquid is $G = 0$. For a rubbery solid, the equilibrium shear modulus is greater than zero but is still many orders of magnitude smaller than the modulus of compression. The ideal-rubber approximation includes the limit that the volume of the system is independent of deformation as long as only one or two dimensions are constrained. This approximation is equivalent to the assertion that the ratio G/K approaches zero.

To simplify the discussion of rubber elasticity, only uniaxial deformation will be considered in this chapter. More complicated strain functions will be considered in the chapter on gels. Consider a uniaxial deformation in the x-direction. It is convenient to define a deformation ratio:

$$\vec{\alpha} = \begin{bmatrix} \alpha_x \\ \alpha_y \\ \alpha_z \end{bmatrix} = \begin{bmatrix} L_x / L_{x0} \\ L_y / L_{y0} \\ L_z / L_{z0} \end{bmatrix}. \tag{4.2}$$

For an ideal rubber, the three components are not independent:

$$\alpha_y = \alpha_z = 1 / \sqrt{\alpha_x} . \tag{4.3}$$

It is then convenient to define a unique deformation variable called the elongation: $\alpha = L_x/L_{x0}$. The thermodynamics of ideal rubber can then be developed as a function of T, α, and the mass of the system, m.

4.2 The thermodynamics of ideal-rubber elasticity

The thermodynamic properties of an ideal rubber can be expressed in terms of the Helmholtz energy, $A(T,\alpha,m)$. The exact differential of the Helmholtz energy for a closed system is then:

$$dA = -S(T,\alpha,m)dT + f(T,\alpha)L_{x0}d\alpha \tag{4.4}$$

where S is the entropy of the sample and f is called the force of extension. Measurements of the force of extension as a function of temperature and elongation have been carried out for many samples of rubber. At constant elongation, the force of extension increases linearly with temperature. The slope of a plot of f against T increases as α increases. A typical plot is shown in Figure 4.1. This result can be understood by thermodynamic analysis.

The force of extension is a thermodynamic function with exact differential:

$$df = \left(\frac{\partial f}{\partial T} \right)_\alpha dT + \left(\frac{\partial f}{\partial \alpha} \right)_T d\alpha . \tag{4.5}$$

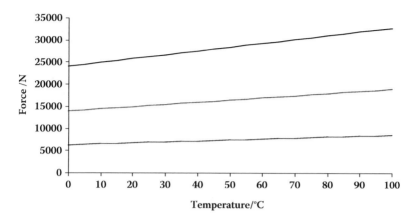

Figure 4.1 The force of retraction of a strip of rubber is plotted linearly against temperature at different values of the extension, α. The three lines correspond to values of the extension equal to 1.1, 1.25, and 1.5. The slope increases as the extension is augmented.

The force of extension can also be expressed as:

$$f\left(T,\alpha\right)=\left(1/L_{x0}\right)\left(\frac{\partial A}{\partial \alpha}\right)_{T}$$

$$f\left(T,\alpha\right)=\left(1/L_{x0}\right)\left[\left(\frac{\partial U}{\partial \alpha}\right)_{T}-T\left(\frac{\partial S}{\partial \alpha}\right)_{T}\right]$$

(4.6)

For an ideal gas, the internal energy, U, is independent of volume. For an ideal rubber, the internal energy is independent of the elongation: $\left(\partial U / \partial \alpha\right)_{T} = 0$. The final necessary identity is obtained from Equation 4.4 by using the Euler relations:

$$\left(\frac{\partial S}{\partial \alpha}\right)_{T} = -L_{x0}\left(\frac{\partial f}{\partial T}\right)_{\alpha}.$$

(4.7)

The elastic equation of state can then be expressed as:

$$f\left(T,\alpha\right)=T\left(\frac{\partial f}{\partial T}\right)_{\alpha}.$$

(4.8)

The force of extension is linearly proportional to temperature, with a slope that depends explicitly on the elongation. The actual value of the slope can be calculated if the entropy change associated with elongation can be derived. The statistical theory of rubber elasticity is presented in Section 4.3. The entropy change, $\Delta S(\alpha)$, is calculated as a function of α. A good discussion of the thermodynamics of rubber elasticity is found in *The Physics of Rubber Elasticity*.[42] Section 4.3 is based on *Principles of Polymer Chemistry*.[15] The understanding of the molecular basis of rubber elasticity is one of the central concepts of polymer science.

4.3   The statistical theory of rubber elasticity

The essential concept involved in the statistical theory of rubber elasticity is that a macroscopic deformation of the whole sample leads to a microscopic deformation of individual polymer chains. The microscopic model of an ideal rubber consists of a three-dimensional network with junction points of known functionality greater than 2. An ideal rubber consists of fully covalent junctions between polymer chains. At short times, high-molecular-weight polymer liquids behave like rubber, but the length of the chains needed to describe the observed elastic behavior is independent of molecular weight and is much shorter than the whole chain. The concept of intrinsic entanglements in uncrosslinked polymer liquids is now well established, but the nature of these restrictions to flow is still unresolved. The following discussion focuses on ideal covalent networks.

The state of the ideal rubber can be specified by the locations of all the junction points, $\{\vec{r}_j\}$, and by the end-to-end vectors for all the chains connecting the junction points, $\{\vec{R}_k\}$. The first postulate of the statistical theory of rubber elasticity is that, in the rest state with no external constraints, the distribution function for the set of chain end-to-end vectors is a Gaussian distribution with a mean-squared end-to-end distance $\langle R^2 \rangle_0$ that is proportional to the molecular weight of the chains between junctions:

$$\wp_0\!\left(\vec{R}\right) = \left(\frac{3}{2\pi\langle R^2 \rangle_0}\right)^{3/2} \exp\!\left(\frac{-3\left(R_x^2 + R_y^2 + R_z^2\right)}{2\langle R^2 \rangle_0}\right). \tag{4.9}$$

There is no preferred direction for the end-to-end vectors in the rest state, and the full set of lengths is represented by the ensemble of chains that constitute the sample of rubber.

The second postulate of the ideal-rubber theory is that, after deformation, the distribution of chain end-to-end vectors is perturbed in exactly the ratio determined by the macroscopic deformation. This assumption is called the principle of **affine deformation**. The distribution of chain end-to-end vectors is now given by:

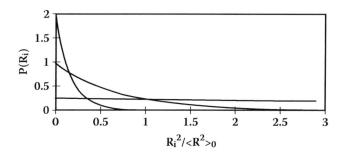

Figure 4.2 Distribution functions for the components of the end-to-end vector in the unperturbed and stretched state, normalized to 1 in the rest state ($\alpha = 4$).

$$\wp_\alpha\left(\vec{R}\right) = \left(\frac{3}{2\pi\left\langle R^2\right\rangle_0}\right)^{3/2} \exp\left(\frac{-3\left(\left(R_x / \alpha\right)^2 + \alpha\left(R_y^2 + R_z^2\right)\right)}{2\left\langle R^2\right\rangle_0}\right). \tag{4.10}$$

In the rest state, the distribution of chain end-to-end vectors is spherically symmetric. In an ideal uniaxial deformation, the distribution of x-, y-, and z-components is cylindrically symmetric. The normalized distribution for the x-component (the stretch direction) is:

$$\wp\left(R_x\right) = \left(\frac{3}{2\pi\alpha^2\left\langle R^2\right\rangle_0}\right)^{1/2} \exp\left(\frac{-3R_x^2}{2\alpha^2\left\langle R^2\right\rangle_0}\right). \tag{4.11}$$

The distribution is now much broader. The corresponding distribution for either the y- or z-component is:

$$\wp\left(R_{y\,or\,z}\right) = \left(\frac{3\alpha}{2\pi\left\langle R^2\right\rangle_0}\right)^{1/2} \exp\left(\frac{-3\alpha R_{y\,or\,z}^2}{2\left\langle R^2\right\rangle_0}\right). \tag{4.12}$$

These two distributions are sharper. A graph of an unperturbed component distribution and two perturbed distributions is shown in Figure 4.2.

The entropy change due to deformation depends explicitly on the two distributions of end-to-end vectors. The exact expression for a system that contains N chains between junctions is:

$$\Delta S(\alpha) = Nk_b \int\limits_{\text{all } \vec{R}} \wp_\alpha(\vec{R}) \ln\left(\frac{\wp_0(\vec{R})}{\wp_\alpha(\vec{R})}\right) d\vec{R} \ . \tag{4.13}$$

Despite the forbidding appearance of the integral, it is easily evaluated with the help of a table of Gaussian integrals. The result is:

$$\Delta S(\alpha) = -\left(Nk_b / 2\right)\left(\alpha^2 + (2/\alpha) - 3\right). \tag{4.14}$$

The force of extension is then given by:

$$f(T,\alpha) = -\left(\frac{T}{L_{x0}}\right)\left(\frac{\partial \Delta S(\alpha)}{\partial \alpha}\right)_T = \left(\frac{Nk_b T}{L_{x0}}\right)\left(\alpha - (1/\alpha^2)\right). \tag{4.15}$$

The force of extension is predicted to be linear in the temperature with a slope that depends on the elongation.

4.4 Thermoelastic inversion point

One of the most remarkable properties of a strip of rubber is that, for a particular fixed length, L_{inv}, the force of extension is independent of temperature:

$$\left(\frac{\partial f}{\partial T}\right)_{L_{\text{inv}}} = 0 \ . \tag{4.16}$$

This phenomenon is called the thermoelastic inversion point because, for longer lengths, the partial derivative is positive and, for shorter lengths, it is negative. The change in the force due to the change in temperature is balanced by the change in the rest length due to the thermal expansion of the rubber. At fixed length, the expansion factor is a function of temperature.

4.5 The force of extension of real rubber

The ideal theory of rubber elasticity explains the linear dependence of the force on the temperature at constant extension, but the dependence of the force on extension at constant temperature is not well represented by Equation 4.15. The change in force with extension, $(\partial f/\partial \alpha)_T$, is usually underestimated in the initial part of the expansion. The effective number of chains in Equation 4.15 appears to exceed the number of network chains determined by the synthesis of the network.

The presence of intrinsic entanglements has been invoked to explain the increased force. As the extension is increased, the observed force approaches the force calculated by Equation 4.15. Apparently, the entanglements are less effective at higher elongations. At even higher extensions, the force rapidly rises with stretching. The explanation of this phenomenon was obtained by examining the X-ray scattering from a strip of rubber during stretching. The rise in force coincides with the development of crystalline order in the sample. The alignment of the chains caused by the stretching raises the melting point of the polymer by increasing the chemical potential of the amorphous phase until it exceeds the chemical potential of the crystalline phase at the stretching temperature. The composite material produced by stretching can be reversibly restored to the relaxed rubber. While substantially more physics must be included to explain the actual behavior of real rubber, the essential character of the force of extension is revealed by the ideal-rubber theory.

chapter five

Structure and properties of polymers in dilute solution

5.1 Introduction

The detailed structure of an isolated polymer molecule in a solution depends on the solvent and the temperature as well as the intramolecular energies discussed in Chapter 2. This chapter presents the principles necessary to understand the solvent and temperature dependence of the global structure. The classical name for this issue is the "excluded volume problem."

The thermodynamic properties of a dilute polymer solution depend on the interaction of polymer molecules with the solvent and with each other. Thermodynamic functions can be expressed in terms of virial expansions in the polymer concentration. The osmotic pressure will be explained in terms of the potential of mean force between pairs of polymer molecules. A detailed discussion of the osmotic second virial coefficient, A_2, will be presented.

One of the most useful techniques for the study of the structure and thermodynamics of dilute polymer solutions is light scattering. The principles of light scattering from an ensemble of solute molecules will be presented and illustrated. The analysis of a Zimm plot will also be explained and discussed.

Polymer molecules are in constant motion in solution. The long-time trajectory of the center of mass is governed by the self-diffusion coefficient, D_s. The molecular theory of the self-diffusion coefficient of a polymer molecule will be presented. The concept of the molecular friction coefficient will be developed. The process of mutual diffusion will then be presented, and an expression for the mutual-diffusion coefficient, D_m, will be derived.

Dynamic light scattering allows the determination of the self-diffusion coefficient and the mutual-diffusion coefficient. In addition, internal conformational fluctuations can be observed for large-chain molecules.

One of the most dramatic properties of a polymer solution is the large increase in viscosity with a small increase in concentration. The molecular theory of solution viscosity in the dilute limit will be presented. The intrinsic viscosity will be derived for the same model used in the discussion of the self-diffusion coefficient.

Measurements of polymer properties in dilute solution are often carried out on samples with a distribution of molecular weights. The properties of molecular-weight distributions and the characteristic averages of the distribution are presented. The relations between the measured quantity and the molecular-weight distribution are derived for light scattering, osmotic pressure, and viscosity.

One of the most useful techniques for measuring the molecular-weight distribution is size-exclusion chromatography (SEC). The principles involved in this technique are also presented.

5.2 The structure of macromolecules in dilute solution

A long-chain molecule in dilute solution can be characterized by a large number of conformations. However, an exact expression for the energy in every conformation is not available. In the rotational isomeric state (RIS) approximation, the energies associated with pairs of local RIS states were considered. When only nearest-neighbor interactions are considered, the mean-squared end-to-end distance can be calculated for a chain of any length. In the random-coil limit, the probability distribution for the end-to-end vector is given by Equation 2.29. However, these theories ignore the energy of interaction between subunits separated by many intervening mers. When the total energy of a single macromolecule and the solvent is considered as a function of the conformation of the chain, the probability distribution for the end-to-end distance depends on the temperature and the solvent.

The measured mean-squared radius of gyration of polystyrene as a function of temperature in Decalin (decahydronaphthalene) is shown in Figure 5.1. There is a large change in the global dimensions as the temperature is changed in dilute solution. While there is no exact solution to the excluded-volume problem, Flory[14] introduced the key approximations that allow a transparent calculation of the change in global dimensions.

In order to discuss the behavior of single chains in solution, an appropriate geometric description must be chosen. Because only global dimensions are being considered, it is convenient to adopt the model of a chain molecule that consists of m subchains that are themselves long enough to exhibit the asymptotic behavior of a random coil. Each subchain is characterized by an end-to-end vector, $\vec{r}_k$, and the total end-to-end vector, $\vec{R}$, is the vector sum of the subchain vectors (Equation 2.26). It is also necessary to specify the location of the center of mass of each subchain, $\vec{R}_k$. The intersubunit vectors, $\vec{R}_{kl} = \vec{R}_k - \vec{R}_l$, are important quantities in the description of the global conformational state of a real-chain molecule in dilute solution.

Now consider a pair of subchains in solution. The Gibbs energy of the solution is a function of the distance between the two subchains, R_{kl}, averaged over all conformations of the pair of subunits and of the associated

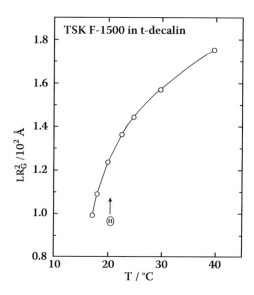

Figure 5.1 The root mean-squared radius of gyration, $\left\langle R_G^2 \right\rangle^{1/2}$, of polystyrene of molecular weight $M = 1.2 \times 10^7$ g/mol in Decalin as a function of temperature. (From Nose, T. and Chu, B., *Macromolecules*, 12, 1122, 1979. With permission.)

solvent. This energy is called the potential of mean force for a pair of subunits, $W(R_{kl})$. The shape and magnitude of this potential energy depends on the solvent and the temperature. The absolute energy is chosen so that the potential of mean force vanishes when the subunits are far apart. For a molecule composed of m subchains, the total energy can be expressed in terms of the constant energy associated with the corresponding Gaussian chain (also called a phantom chain because intersubunit energies are ignored) and the energies of interaction of all pairs of subunits:

$$E\left(\left\{\vec{R}_k\right\}\right) = W_0 + \sum_{k<l} W\left(R_{kl}\right) \tag{5.1}$$

where the sum is over all unique unlike pairs of subunits. The probability of any particular global conformation is then proportional to its Boltzmann factor:

$$\wp\left(\left\{\vec{R}_k\right\}\right) \propto \exp\left(\frac{-E\left(\left\{\vec{R}_k\right\}\right)}{k_b T}\right) \tag{5.2}$$

The interaction of any two subchains can be characterized by an integral called the excluded volume for a pair of subchains:

$$V_e = (1/2) \int_0^\infty \left(1 - \exp\left(\frac{-W(R_{kl})}{k_b T}\right)\right) 4\pi R_{kl}^2 \, dR_{kl} \, . \tag{5.3}$$

If there is no potential of mean force between subunits, the excluded-volume integral must vanish. The form of $W(R_{kl})$ is complicated for a real macromolecule; it is possible for two subunits to have the same location for the center of mass. The value of $W(0)$ is generally positive and finite. The potential of mean force is defined to be zero when the two subchains are far apart. The value of the potential of mean force at separations comparable to twice the radius of gyration of the subchains depends strongly on the solvent. If the solvent is chemically similar to the mers of the macromolecule (a "good" solvent), the potential of mean force reflects primarily the decrease in entropy caused by the overlap of the two subchains. In the good-solvent limit, the excluded volume is eight times the volume of a single subchain. Under these conditions, the chain will expand its average radius of gyration to minimize the number of intersubunit interactions. It is convenient to define an expansion factor $\alpha^2 = \langle R_G^2 \rangle / \langle R_G^2 \rangle_0$, where the subscript 0 denotes the value for the corresponding phantom chain. For a Gaussian chain, the mean-squared radius of gyration is proportional to the number of subunits, m (Equation 2.27). In the good-solvent limit, it is observed experimentally that $\langle R_G^2 \rangle$ scales as $m^{6/5}$. The expansion factor depends on the chain length.

If the energy of interaction of two mers is more favorable than the interaction of a solvent molecule and a mer, the potential of mean force for weak overlap of two subchains will be negative. The value of the excluded volume will depend on the temperature through the denominator in the exponential term of Equation 5.3. It is possible to have positive excluded volumes in one range of temperature and negative excluded volumes in a different range. The intersubunit excluded volume vanishes at a particular temperature. Because the potential of mean force depends on both the polymer and the solvent, it is even possible to observe more than one temperature at which $V_e = 0$. The existence of a compensation point where the intersubunit repulsion at small separations is balanced by the attraction at larger distances is exactly comparable to the Boyle temperature of a real gas. It might then be supposed that, at the compensation temperature for the intersubunit excluded volume, the probability of any global conformation would be the same as that for a phantom chain. There are reasons to believe that this is not the case. The concentration of subchains in a real macromolecule is too high to truncate the energy expression (Equation 5.1) with terms involving only pairs of subunits. There are additional energies associated with the interactions of three subchains that are not comprehended by the sum of the three pairwise interactions. These energies are usually positive. It is necessary for the excluded volume associated with pairs of subunits to be slightly negative in order for the overall mean-squared radius of gyration to adopt its phantom-chain value.

At temperatures below the compensation temperature, the chain contin-ues to collapse until it reaches a density high enough to enter the globule state. In the globule state, the mean-squared radius of gyration is propor-tional to the quantity $m^{2/3}$. While the polymer volume fraction in the globule state can still be small, the average subunit density is much more uniform than in the random-coil state. The crossover from the expanded random-coil state to the collapsed globule state with temperature is an example of an intramolecular phase transition.

5.3 Flory theory of chain expansion

The Flory theory of chain expansion results in an explicit expression for the expansion factor. The key concept is that the energy of interaction of all the pairs of subchains depends on the distribution of subchains relative to the center of mass. The theory starts with a consideration of the probability distribution for the radius of gyration of a long-chain molecule. The exact probability distribution for the radius of gyration of a phantom chain is complicated, but a good approximation was chosen by Flory:

$$\wp_0\left(R_G\right)=\left(\frac{3}{2\pi\left\langle R_G^2\right\rangle_0}\right)^{3/2}\exp\left(\frac{-3R_G^2}{2\left\langle R_G^2\right\rangle_0}\right)4\pi R_G^2 . \tag{5.4}$$

To consider the effect of intersubunit interactions, Flory defined an energy for the chain as a function of the radius of gyration, averaged over all global conformations with the same radius of gyration, $W(R_G)$. The prob-ability distribution for the radius of gyration of a real chain is then postulated to be:

$$\wp\left(R_G\right)=\frac{\wp_0\left(R_G\right)\exp\left(\dfrac{-W\left(R_G\right)}{k_bT}\right)}{\displaystyle\int_{\text{all }R_G}\wp_0\left(R_G\right)\exp\left(\dfrac{-W\left(R_G\right)}{k_bT}\right)dR_G} . \tag{5.5}$$

The energy function, $W(R_G)$, was calculated by assuming that the local energy of interaction is proportional to the square of the local subunit density. The subunit density was obtained by assuming that the subchains were distributed relative to the center of mass in a Gaussian distribution. This model is called the smoothed-density model and will be used many times in this chapter. The number density of subunits at a location $\vec{S}$ relative to the center of mass in a macromolecule of overall radius of gyration R_G is given by:

$$\rho\left(\vec{S}, R_G\right) = m\left(\frac{3}{2\pi R_G^2}\right)^{3/2} \exp\left(\frac{-3S^2}{2R_G^2}\right). \tag{5.6}$$

The potential is then proportional to the integral of the square of the subunit density. The result can be expressed in the form:

$$W\left(R_G\right) = Ck_b T\left(\frac{V_e\, m^2}{R_G^3}\right) \tag{5.7}$$

where C is a numerical constant. The intersubunit energy function depends on the excluded volume.

The chain-expansion factor can be calculated by evaluating the mean-squared radius of gyration using Equation 5.5. Flory obtained an equation that can be written as:

$$\alpha^5 - \alpha^3 = C\left(\frac{V_e\, m^2}{\left\langle R_G^2\right\rangle_0^{3/2}}\right). \tag{5.8}$$

The phantom-chain quantity in the denominator is proportional to $m^{3/2}$. When the chain-expansion factor is large, the squared expansion factor is proportional to $m^{1/5}$. This is what is observed for long chains in good solvents.

5.4   Thermodynamics of two component solutions

A two-component solution is characterized by a Gibbs energy $G\left(T, P, N_1, N_2\right)$, where N_1 is the number of molecules of component 1, which will be called the solvent, and N_2 is the number of molecules of component 2, which will be called the solute. It is convenient to express the Gibbs energy as:

$$G(T, P, N_1, N_2) = G_1^\circ\left(T, P, N_1\right) + G_2^\circ\left(T, P, N_2\right) + \Delta G_{mix}\left(T, P, N_1, N_2\right) \tag{5.9}$$

where G_1° is the Gibbs energy of the pure solvent, G_2° is the Gibbs energy of the pure solute, and ΔG_{mix} is the Gibbs energy of mixing. All thermodynamic properties of the solution can be calculated from the Gibbs energy function.

The chemical potential of the solvent in the solution is defined as:

$$\mu_1 = \left(\frac{\partial G}{\partial N_1}\right)_{T,P,N_2} = \mu_1^\circ\left(T, P\right) + \left(\frac{\partial \Delta G_{mix}}{\partial N_1}\right)_{T,P,N_2} \tag{5.10}$$

where μ_1° is the chemical potential of the pure solvent. The chemical potential can be expressed as a function of temperature, pressure, and composition. Changes in the chemical potential of the solvent can then be expressed as:

$$d\mu_1 = -s_1 dT + v_1 dP + \left(\frac{\partial \mu_1}{\partial x_2}\right)_{T,P} dx_2 \qquad (5.11)$$

where $s_1 = \left(\dfrac{\partial \mu_1}{\partial T}\right)_{P,x_2}$ is called the partial entropy of the solvent; $v_1 =$

$\left(\dfrac{\partial \mu_1}{\partial P}\right)_{T,x_2}$ is called the partial volume of the solvent; and $x_2 = \dfrac{N_2}{N_1 + N_2}$ is

the mole fraction of the solute. It is also conventional to define an activity function, a_1, as:

$$\mu_1(T,P,x_2) = \mu_1^\circ(T,P) + k_b T \ln a_1(T,P,x_2). \qquad (5.12)$$

5.5 Osmotic pressure

The chemical potential of the solvent is lowered by an increase in temperature because the partial entropy of the solvent must be positive. The chemical potential is increased by an increase in pressure because the partial volume of the solvent must be positive. In a dilute solution, the chemical potential of the solvent must be lowered by an increase in the solute concentration because the quantity $(\partial \mu_1/\partial x_2)_{T,P}$ is negative. In a more concentrated solution, the value of this partial derivative can have either sign. The decrease in the chemical potential due to the addition of solute can be balanced by a change in temperature or pressure. The melting-point depression is determined by the temperature change necessary to achieve a solvent chemical potential equal to the pure component solid. The osmotic pressure is determined by the pressure change necessary to achieve a solvent chemical potential equal to $\mu_1^\circ(T,P)$. The formal expression of this statement is:

$$\mu_1(T,P+\pi,x_2) = \mu_1^\circ(T,P). \qquad (5.13)$$

It is also possible to calculate the change in the chemical potential of the solvent due to the increase in pressure:

$$\mu_1(T,P+\pi,x_2) - \mu_1(T,P,x_2) = \int_P^{P+\pi} v_1(T,P,x_2)\,dP. \qquad (5.14)$$

Although the partial volume of the solvent is formally a function of temperature, pressure, and composition, it is customary to ignore the small dependence of this quantity on these variables in a dilute solution and to treat the partial volume as equal to the pure component value, v_1°. The integral is then easily carried out. When Equation 5.13 is invoked, the osmotic pressure can be expressed as:

$$\pi\left(T,x_2\right) = -\frac{\left(\mu_1\left(T,P,x_2\right) - \mu_1^{\circ}\left(T,P\right)\right)}{v_1^{\circ}}. \tag{5.15}$$

The osmotic pressure is a direct measure of the change in chemical potential of the solvent due to the addition of solute. It can be measured by several experimental techniques. The osmotic pressure can then be compared with calculations of the change in chemical potential upon dilution.

The osmotic pressure of an ideal solution can be calculated from the phenomenological expression for the chemical potential (Equation 5.12). The pressure dependence of the activity is ignored:

$$\pi\left(T,x_2\right) = -\left(k_b T / v_1^{\circ}\right)\ln a_1\left(T,x_2\right). \tag{5.16}$$

A typical definition of an ideal solution is one that obeys Raoult's Law. This definition is asymptotically correct in dilute solution. The osmotic pressure can then be expressed as:

$$\pi\left(T,x_2\right) = -\left(k_b T / v_1^{\circ}\right)\ln\left(1-x_2\right). \tag{5.17}$$

In the limit of a very dilute solution, Equation 5.17 becomes:

$$\pi = \frac{N_2 k_b T}{V} \tag{5.18}$$

which is the van't Hoff Law. Even a polymer solution obeys the van't Hoff Law in the limit of a very dilute solution. It is customary to express the polymer concentration in units of mass per unit volume, c. In these units, Equation 5.18 becomes:

$$\pi = \frac{cRT}{M}. \tag{5.19}$$

The osmotic pressure of an actual dilute polymer solution is usually written as:

$$\pi(T,c) = \left(\frac{cRT}{M}\right)\left[1 + A_2 Mc + \cdots\right]$$
(5.20)

where A_2 is called the osmotic second virial coefficient. The value of A_2 depends on temperature and, for a particular polymer-solvent pair, can be positive in one temperature range and negative in another range. There is a temperature at which the osmotic second virial coefficient vanishes. This phenomenon is also similar to the Boyle point in a real gas and arises from the same effect considered in Equation 5.2. The temperature at which $A_2 = 0$ is called the Flory temperature. The theory of the osmotic second virial coefficient developed by Flory is described in the following section.

5.6 Flory theory of the second virial coefficient

The osmotic second virial coefficient, A_2, is determined by the interaction of two polymer molecules with each other and with the solvent. This interaction can be expressed in terms of a potential of mean force, $U(R_{12})$, where R_{12} is the distance between the centers of mass of the two macromolecules. In terms of the potential of mean force, the osmotic second virial coefficient is given by:

$$A_2(T,M) = \left(\frac{N_A}{2M^2}\right)\int_0^\infty \left(1 - \exp\left(\frac{-U(R_{12})}{k_b T}\right)\right) 4\pi R_{12}^2 dR_{12} .$$
(5.21)

The integral is twice the excluded volume for a pair of whole polymer molecules. The osmotic second virial coefficient is a direct measure of the excluded volume of the polymer chains. If the chains are modeled as hard spheres with radius R, the second virial coefficient is:

$$A_{2,hs}(T,M,R) = \left(\frac{16\pi R^3 N_A}{3M^2}\right).$$
(5.22)

For globular protein molecules in a solution of high ionic strength at the isoelectric point, the second virial coefficient is well described by Equation 5.22. Since the volume of a globular protein is proportional to its molecular weight, the second virial coefficient decreases as $1/M$.

An approximate potential of mean force for two polymer coils in a good solvent was developed by Flory and Krigbaum.[14] It is based on the same smoothed density approximation presented in Section 5.3. The Flory–Krig-baum potential can be expressed as:

$$U\left(R_{12}\right) = 2k_b T m^2 V_e \left(\frac{3}{4\pi \left\langle R_G^2 \right\rangle}\right)^{3/2} \exp\left(-\frac{3R_{12}^2}{4\left\langle R_G^2 \right\rangle}\right). \tag{5.23}$$

The potential is a Gaussian with range determined by the mean-squared radius of gyration and strength determined by the intersubunit excluded volume and the square of the number of subchains. Direct evaluation of the integral (Equation 5.21) then leads to the lowest-order solution for the second virial coefficient:

$$A_2\left(T, M\right) = \left(\frac{N_A m^2 V_e}{M^2}\right). \tag{5.24}$$

The intermolecular excluded volume is given by the intersubunit excluded volume times the square of the number of subchains. In this approximation, the second virial coefficient for Gaussian chains is independent of molecular weight. Actual measurements on high-molecular-weight polymers in good solvents reveal that A_2 decreases slowly with molecular weight. The full solution for the second virial coefficient can be expressed as a product of the factor in Equation 5.24 and a function $h(z)$, where z is proportional to the intersubunit excluded volume times $m^{1/2}$. The decrease in $h(z)$ with increasing chain length accounts for the observed decrease in the second virial coefficient. Yamakawa[44] provides an extensive discussion of the function $h(z)$.

A simple concept that predicts the correct behavior for the osmotic second virial coefficient in the high-molecular-weight limit in a good solvent is due to deGennes.[9] High-molecular-weight polymers in a good solvent do not overlap much because the decrease in entropy caused by the overlap is very large. The excluded volume for two random coils then becomes proportional to the pervaded volume of the chain molecule, $V_p = \left(4\pi / 3\right)\left\langle R_G^2 \right\rangle^{3/2}$. Within the good-solvent limit, the pervaded volume scales as $M^{9/5}$. The second virial coefficient is then proportional to $M^{-1/5} = M^{9/5}/M^2$, in good agreement with experimental observations.

While the Flory–Krigbaum potential is a qualitatively reasonable form for polymers in a good solvent, it completely misrepresents the shape of the potential of mean force near the Flory temperature. The potential of mean force is positive and repulsive for small values of the center-of-mass separation, just as it was for the overlap of two subchains. For the overall excluded volume to vanish, the potential of mean force must be negative for small degrees of overlap. The Flory compensation temperature is a balance between the inherent repulsion of two polymer coils at high degrees of overlap and the attraction of two mers relative to the solvent.

5.7   Light scattering from two-component solutions

In Chapter 3, the scattering function for a single macromolecule was introduced. The scattered intensity was shown to be a function of the magnitude of the scattering vector q. In this section, the value of q is restricted to the very low range. At each concentration, there is a characteristic distance, L_c, between solute molecules that is proportional to $c^{-1/3}$. When the quantity qL_c $<< 1$, it is convenient to treat the solution as a continuum characterized by a local concentration, $c(\vec{r}, t)$, that fluctuates in both position and time. If the refractive index of the solution, n, depends on the concentration of the solute, light will be scattered by concentration fluctuations. The theory of light scattering by concentration fluctuations was first derived by Einstein.

The results of a light-scattering experiment are usually expressed in terms of a scattering cross section called the Rayleigh ratio, $R(q)$. It is defined as:

$$R(q) = \frac{I(q)L^2}{V_s I_0}$$
(5.25)

where $I(q)$ is the intensity detected at a particular angle (q); L is the distance from the scattering volume, V_s, to the detector; and I_0 is the incident intensity. The intensity observed from the pure solvent is subtracted from the total intensity to produce an excess Rayleigh ratio $\Delta R(q) = R(q) - R^o(q)$. The Einstein theory for light scattering from concentration fluctuations can then be expressed as:

$$\Delta R(q) = \left(\frac{2\pi n^2 (dn/dc)^2 V_s}{\lambda^4} \right) \langle \Delta c(q) \Delta c(q) \rangle$$
(5.26)

where (dn/dc) is the refractive increment, and:

$$\Delta c(q) = \int_{\text{all } \vec{r}} \left(c(\vec{r}) - \langle c \rangle \right) \exp\left(i\vec{q} \cdot \vec{r} \right) d\vec{r}$$
(5.27)

is the spatial Fourier transform of the instantaneous concentration fluctuations. The brackets denote an average over time of the squared concentration fluctuation with spatial Fourier component q. In the present discussion, the limit $q = 0$ will be taken. This is called the thermodynamic limit. The mean-squared concentration fluctuation in a two-component system is given by:

$$\left\langle \Delta c^2 \right\rangle = -\frac{c v_1^\circ N_A k_b T}{V_s \left(\partial \mu_1 / \partial c \right)_{T,P}} . \tag{5.28}$$

The partial derivative can be related to the osmotic pressure. The excess Rayleigh ratio for concentration fluctuations can then be expressed as:

$$\Delta R(c) = \frac{2\pi^2 n^2 \left(dn / dc \right)^2 c N_A k_b T}{\lambda^4 \left(\partial \pi / \partial c \right)_{P,T}} . \tag{5.29}$$

To focus on the variables that change with the solution, the excess Rayleigh ratio is often written as:

$$\Delta R(c) = KcMS(c) \tag{5.30}$$

where $S(c) = \dfrac{RT}{M \left(\partial \pi / \partial c \right)_{P,T}}$ is called the solution structure factor. Within the dilute-solution limit, the structure factor is given by:

$$S(c \to 0) = \frac{1}{1 + 2A_2 Mc + \cdots} . \tag{5.31}$$

In the thermodynamic limit ($q \to 0$), measurement of the solution structure factor is a good method for obtaining the osmotic second virial coefficient. Measurements of the excess Rayleigh ratio as a function of concentration yield the molecular weight of the solute within this limit. The usual procedure is to plot the quantity $Kc/\Delta R(c)$ against concentration:

$$\frac{Kc}{\Delta R(c)} = \frac{1}{M} + 2A_2 c + \cdots . \tag{5.32}$$

The molecular weight is obtained from the intercept and the second virial coefficient from the slope.

5.8   Flory theory of light scattering in dilute solution

The excess Rayleigh ratio can also be calculated for the general case of a dilute solution at all scattering angles.[13] The scattering function at finite scattering vector is defined as:

$$S(q,c) = (1/N)(1/m^2)\left\langle \sum_{i,j}\sum_{k,l}\exp\left(i\vec{q}\cdot\left(\vec{r}_{ik}-\vec{r}_{jl}\right)\right)\right\rangle \tag{5.33}$$

where N is the number of macromolecules in the scattering volume, m is the number of subchains for each polymer; $\vec{r}_{ik}$ is the location of the ith subchain on the kth molecule; the summation is over all pairs of subchains in the scattering volume; and the brackets denote an average over all conformations of the system. The excess Rayleigh ratio is $\Delta R(q,c) = KcMS(q,c)$.

To evaluate this expression, it is convenient to express the location of each subchain in terms of the location of the center of mass of each macromolecule, $\vec{R}_k$, and the location of the subchain relative to the center of mass, $\vec{s}_{ik}$: $\vec{r}_{ik} = \vec{R}_k + \vec{s}_{ik}$. The scattering function can then be expressed in terms of an average over the locations of the centers of mass of different molecules and an average over the subchains on the same molecule:

$$S(q,c) = (1/N)\left\langle \sum_{k,l}\exp\left(i\vec{q}\cdot\vec{R}_{kl}\right)\right\rangle(1/m^2)\left\langle \sum_{i,j}\exp\left(i\vec{q}\cdot\left(\vec{s}_i-\vec{s}_j\right)\right)\right\rangle \tag{5.34}$$

where $\vec{R}_{kl} = \vec{R}_k - \vec{R}_l$ and the sum is over all pairs of solute molecules.

The difference in the moment of inertia vectors is equal to the difference in the absolute location vectors of the subchains: $\vec{s}_i - \vec{s}_j = \vec{r}_{ij}$. The second average is then equal to the scattering function for a single molecule, $S_{mol}(q)$ (Equation 3.7). The total scattering function is then expressed as the product of the center-of-mass structure factor and the intramolecular structure factor:

$$S(q,c) = S_{cm}(q,c)S_{mol}(q). \tag{5.35}$$

The center-of-mass structure factor can be expressed in terms of the pair correlation function, $g(R_{kl})$, for the centers of mass of different molecules. Within the limit of infinite dilution, the pair correlation is defined exactly in terms of the pair potential of mean force:

$$g(R_{kl},c\rightarrow 0) = \exp\left(\frac{-U(R_{kl})}{k_bT}\right). \tag{5.36}$$

At higher concentrations, the effect of the interactions of three molecules must be taken into account. The center-of-mass structure factor is given by:

$$S_{cm}(q,c) = 1 - (N_2/V)\int_{V_s}\exp\left(i\vec{q}\cdot\vec{R}_{kl}\right)\left(1-g(R_{kl})\right)dR_{kl} \tag{5.37}$$

where the integration is over the scattering volume. Within the thermodynamic limit, it is seen that the center-of-mass structure factor within the dilute solution limit is equal to:

$$S_{cm}(q=0, c \to 0) = 1 - 2A_2 Mc \qquad (5.38)$$

which is the low-concentration expansion of Equation 5.31. At finite scattering vector, the standard approximation of Equation 5.37 within the dilute limit is:

$$S_{cm}(q, c \to 0) = 1 - 2A_2 Mc S_{mol}(q). \qquad (5.39)$$

The quantity $Kc/\Delta R$ can then be expressed as:

$$\frac{Kc}{\Delta R(q, c \to 0)} = \frac{1}{M}\left[\frac{1}{S_{mol}(q)} + 2A_2 Mc + \cdots \right]. \qquad (5.40)$$

Within the low q limit, the intramolecular structure factor is as given in Equation 3.9, and Equation 5.40 becomes:

$$\frac{Kc}{\Delta R(q \to 0, c \to 0)} = \frac{1}{M}\left[1 + \frac{q^2 \langle R_G^2 \rangle}{3} + 2A_2 Mc + \cdots \right]. \qquad (5.41)$$

This theoretical form is the basis of the Zimm plot that yields the molecular weight, the mean-squared radius of gyration, and the osmotic second virial coefficient from measurements of the excess light scattering as a function of angle and concentration. A typical Zimm plot is shown in Figure 5.2.

The reduced experimental quantity, $Kc / \Delta R(q, c)$, is extrapolated to infinite dilution for each value of q, and the mean-squared radius of gyration is obtained by fitting this limiting line to Equation 3.9. The data are extrapolated to $q = 0$ at each concentration, and the molecular weight and osmotic second virial coefficient are obtained from the slope and intercept. To present all the data on the same graph, it is traditional to plot the reduced experimental quantity against the compound variable, $q^2 + \alpha c$, where the scaling factor α is chosen to give a pleasing plot. It is not necessary to make a Zimm plot to carry out a Zimm analysis. It is possible to carry out a least-squares analysis with the three experimental quantities as unknowns and to use all the data at once to constrain the results. This is the normal technique for analyzing large amounts of data. A thorough discussion of such data is found in Min et al.[28] A more detailed discussion of the calculation of the cen-

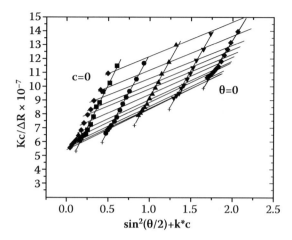

Figure 5.2 Plot of reduced scattered intensities, $Kc/\Delta R$, against the compound variable $\sin^2(\theta/2) + kc$. Analysis of these data yields the molecular weight M, the osmotic second virial coefficient A_2, and the mean-squared radius of gyration $\left\langle R_G^2 \right\rangle$.

ter-of-mass structure factor in dilute solution using the Flory–Bueche theory and the Flory–Krigbaum potential is found in Kim et al.[23]

5.9   Diffusion of particles in solution

Molecules in solution are in constant motion. The trajectory of the center of mass of a macromolecule in solution can be described in terms of a vector, $\vec{R}(t)$. If the origin of coordinates is chosen as the location of the particle at $t = 0$, the squared displacement is given by $\vec{R} \cdot \vec{R}$. If the trajectory is followed many times for some elapsed time τ, the mean-squared displacement is found to be proportional to the elapsed time. The Brownian diffusion law is

$$\left\langle R^2 \right\rangle = 6D_s\tau \tag{5.42}$$

where D_s is called the self-diffusion coefficient. If the particle is a sphere of radius R, the self-diffusion coefficient is inversely proportional to R and directly proportional to the temperature of the solution. It is also inversely proportional to the viscosity of the solution. The explicit formula for the self-diffusion coefficient can be calculated by the method of Brownian dynamics.

The solvent is treated as a viscous continuum of viscosity η and temperature T. A particle in such a solution is subject to three types of forces:

1.   Particle motion is resisted by a viscous force equal to $-\zeta\vec{v}$, where $\vec{v}$ is the particle velocity relative to the solvent, and ζ is called the

friction coefficient. For a sphere of radius R, the friction coefficient was derived by Stokes:

$$\zeta = 6\pi\eta R .$$

(5.43)

2. External forces such as gravity or electric or magnetic fields.
3. A random force, $\vec{A}(\vec{r},t)$, that is due to the motions of the solvent molecules and that varies with both location and time. The random force is uncorrelated in time at the same location and uncorrelated with position at all times.

The trajectory of a particle can then be calculated in terms of the Langevin equation:

$$m\frac{d\vec{v}}{dt} = -\zeta\vec{v} + \sum_i \vec{F}_i + \vec{A}(\vec{r},t)$$

(5.44)

where m is the particle mass, and the sum is over all the external forces. This equation will be used to derive all the properties of a particle undergoing Brownian motion in solution.

Suppose the particle is subject to a gravitational force, $F_g = mg(1-\bar{v}\rho)$, where g is the gravitation constant, $\bar{v}$ is the partial specific volume of the solute particle, and ρ is the density of the solvent. The particle will move in the solution at a rate determined by the gravitational force and the frictional resistance. It will accelerate until the two forces balance and a terminal average velocity is attained. The instantaneous velocity will continue to fluctuate due to the random force, but a well-defined average terminal velocity is observed. The form of Equation 5.44 appropriate for this situation is:

$$0 = m\frac{d\langle\vec{v}\rangle}{dt} = -\zeta\langle\vec{v}\rangle - mg(1-\bar{v}\rho)\hat{z}$$

(5.45)

since $\langle\vec{A}(t)\rangle = 0$. The terminal velocity is then

$$\langle\vec{v}\rangle = -\left(\frac{mg}{\zeta}\right)(1-\bar{v}\rho)\hat{z} .$$

(5.46)

The mean-squared displacement of a particle in a solution subject to no external forces can be expressed as:

$$\langle R^2 \rangle = 6\left(\frac{k_b T}{\zeta}\right)\tau ;$$

(5.47)

within the limit of long elapsed times. This means that the self-diffusion coefficient of a particle in solution is:

$$D_s = \frac{k_b T}{\zeta} .$$
(5.48)

If the particle is spherical, the friction coefficient is given by Equation 5.43. For particles of arbitrary shape, it is customary to define a hydrodynamic radius in terms of the measured friction coefficient:

$$\zeta = 6\pi\eta R_H .$$
(5.49)

5.10   Kirkwood theory[44] of macromolecular friction

The theoretical approach to calculating the friction coefficient, f, of a macromolecule is based on the derivation of the force exerted by the particle on the fluid when it is moving with a center-of-mass velocity, $\vec{u}$: $\vec{F} = f\vec{u}$. The macromolecule is divided into m subchains, with subchain friction coefficient ζ. The total force is obtained as the vector sum of the forces on the subchains:

$$\vec{F} = \sum_{i=1}^{m} \langle \vec{F}_i \rangle ;$$
(5.50)

where the brackets denote an average over all conformations of the molecule. Two further approximations are employed in the calculation:

1. All the subchains move with the center-of-mass velocity.
2. The molecule adopts the equilibrium distribution of subchain conformations.

The solvent is characterized by a velocity field, $\vec{v}(\vec{r}, t)$. The force exerted by the ith subchain on the fluid is given by:

$$\vec{F}_i = \zeta(\vec{u}_i - \vec{v}_i)$$
(5.51)

where $\vec{v}_i$ is the value of the fluid velocity at the location of the center of mass of subchain i, and $\vec{u}_i$ is the velocity of the center of mass of the subchain. The subtle part of this problem arises in the calculation of the local fluid velocity, because the motion of the other $m - 1$ subchains perturbs the local fluid velocity. The velocity perturbation at another location due to a force exerted at the origin is given by:

$$\vec{v}'\left(\vec{r}\right) = \underline{\underline{T}}\vec{F} \tag{5.52}$$

where

$$\underline{\underline{T}} = \left(\frac{1}{8\pi\eta_0 r}\right)\left(1 + \frac{\vec{r}\vec{r}}{r^2}\right) \tag{5.53}$$

is called the Oseen interaction tensor.

The perturbation of the fluid velocity at one of the subchains is the sum of the perturbations from all the other subchains. The local fluid velocity is then:

$$\vec{v}_i = \vec{v}_i^0 + \sum_{i \neq j}^{m} \underline{\underline{T}}_{ij}\vec{F}_j \,. \tag{5.54}$$

There are then m coupled equations for the local fluid velocity and the force exerted by each subchain:

$$\vec{F}_i = \zeta\left(\vec{u}_i - \vec{v}_i^0\right) - \zeta\sum_{j \neq i}^{m} \underline{\underline{T}}_{ij}\vec{F}_j \,. \tag{5.55}$$

This is the fundamental equation derived by Kirkwood and Riseman to treat both macromolecular friction and intrinsic viscosity.

For the derivation of the friction coefficient, the unperturbed fluid velocity is zero, and all subchains move with the center-of-mass velocity. The average force exerted by the ith subchain is then:

$$\left\langle\vec{F}_i\right\rangle = \zeta\vec{u} - \zeta\sum_{j \neq i}^{m} \left\langle\underline{\underline{T}}_{ij}\vec{F}_j\right\rangle \,. \tag{5.56}$$

The calculation of the average of the product of the Oseen tensor and the force exerted by the jth subchain is exceedingly difficult. The standard approximation is to carry out the average over each quantity separately and multiply the results. The average Oseen tensor is:

$$\left\langle\underline{\underline{T}}_{ij}\left(\vec{R}_{ij}\right)\right\rangle = \left(\frac{1}{6\pi\eta_0}\right)\left\langle\frac{1}{R_{ij}}\right\rangle \,. \tag{5.57}$$

The average Oseen tensor is a scalar function of the intersubchain distance R_{ij}. The average over the intersubchain tensor function gives a numerical constant of $4/3$, due to the spherically symmetric nature of $\wp\left(\vec{R}_{ij}\right)$. The force exerted by a subchain can then be expressed as:

$$\left\langle \vec{F}_i \right\rangle = \zeta \vec{u} - \left(\frac{\zeta}{6\pi\eta_0} \right) \sum_{j \neq i}^{m} \left\langle \frac{1}{R_{ij}} \right\rangle \left\langle \vec{F}_j \right\rangle . \tag{5.58}$$

A thorough discussion of the solution to this set of coupled equations is beyond the scope of this book, but the result can be written in the simple form:

$$f = \frac{m\zeta}{1 + (8/3)X} \tag{5.59}$$

where X is called the hydrodynamic draining parameter. If $X = 0$, the macromolecular friction coefficient is simply the sum of the subchain friction coefficients. If $X > 0$, the polymeric friction coefficient is *smaller* than the corresponding free draining ($X = 0$) case. The parameter X is given by:

$$X = \frac{m^{1/2}\zeta}{\left(6\pi^3\right)^{1/2} \eta_0 \left\langle R_{12}^2 \right\rangle^{1/2}} \tag{5.60}$$

where $\left\langle R_{12}^2 \right\rangle$ is the mean-squared distance between the centers of mass of adjacent subchains. Within the limit of large X, the macromolecular friction coefficient is independent of the subchain friction coefficient!

$$f = \left(\frac{9\pi^{3/2}}{4} \right) \eta_0 \left\langle R_G^2 \right\rangle^{1/2} . \tag{5.61}$$

Within this limit, called the nondraining chain, the hydrodynamic radius is proportional to the root mean-squared radius of gyration:

$$R_H = \left(\frac{3\pi^{1/2}}{8} \right) \left\langle R_G^2 \right\rangle^{1/2} . \tag{5.62}$$

Attaining the nondraining limit is usually accomplished by using high-molecular-weight polymers. The cooperative effect of the large number of subchains entrains all the solvent within the coil. Measurements of the macromolecular friction coefficient for a random coil are then interpreted in terms of the mean-squared radius of gyration. A more detailed discussion of the macromolecular friction is found in the monograph by Yamakawa.[44]

5.11   Concentration fluctuations and mutual diffusion

The decay of concentration fluctuations in a two-component mixture is described by Fick's Law. Consider the solution to be characterized by a local concentration, $c(\vec{r},t)$. The time rate of change of the concentration at position $\vec{r}$ is governed by the spatial gradient of the concentration flux vector, $\vec{J}_c(\vec{r},t)$:

$$\left(\frac{\partial c(\vec{r},t)}{\partial t}\right) = -\vec{\nabla}\cdot\vec{J}_c(\vec{r},t) . \tag{5.63}$$

The concentration-flux vector is proportional to the concentration gradient:

$$\vec{J}_c(\vec{r},t) = -D_m\vec{\nabla}c(\vec{r},t) \tag{5.64}$$

where D_m is the mutual-diffusion coefficient.

The derivation of a microscopic expression for the mutual-diffusion coefficient uses Equation 5.44. The relevant force is the osmotic force due to a gradient in the chemical potential of the solution:

$$\vec{F}_\mu(\vec{r},t) = -\vec{\nabla}\mu_2(\vec{r},t) = -\left(\frac{\partial\mu_2}{\partial c}\right)_{T,P}\vec{\nabla}c(\vec{r},t) . \tag{5.65}$$

Under steady-state conditions, the concentration flux vector is:

$$\vec{J}_c(\vec{r},t) = -c\left(\frac{1}{f}\right)\left(\frac{\partial\mu_2}{\partial c}\right)_{P,T}\vec{\nabla}c(\vec{r},t) . \tag{5.66}$$

The mutual-diffusion coefficient is then identified by inspection. It is customary to convert the partial derivative to an expression involving the osmotic pressure:

$$\left(\frac{\partial\mu_2}{\partial c}\right)_{P,T} = \left(\frac{M}{N_A c}\right)(1-\bar{v}c)\left(\frac{\partial\pi}{\partial c}\right)_{P,T} . \tag{5.67}$$

The mutual-diffusion coefficient can then be written as:

$$D_m = \left(\frac{M}{N_A f}\right)(1 - \bar{v}c)\left(\frac{\partial \pi}{\partial c}\right)_{P,T}. \tag{5.68}$$

In dilute solution, the mutual-diffusion coefficient can be expressed as:

$$D_m(c \to 0) = \left(\frac{k_b T}{f}\right)(1 - \bar{v}c)(1 + 2A_2 Mc + \cdots). \tag{5.69}$$

The macromolecular friction coefficient in dilute solution can be written as a virial expansion:

$$f = (6\pi \eta_0 R_H)(1 + k_f c + \cdots) = f_0(1 + k_f c + \cdots). \tag{5.70}$$

The mutual-diffusion coefficient in dilute solution can then be expressed as:

$$D_m(c \to 0) = D_0\left(1 + (2A_2 M - k_f - \bar{v})c + \cdots\right) \tag{5.71}$$

where D_0 is the self-diffusion coefficient at infinite dilution. The mutual-diffusion coefficient can be either greater or less than the self-diffusion coefficient in dilute solution. At the Flory temperature, the mutual-diffusion coefficient falls with concentration. Within the good-solvent limit for high-molecular-weight polymers, the mutual-diffusion coefficient can greatly exceed the self-diffusion coefficient.

The best way to measure the mutual-diffusion coefficient of a dilute polymer solution is light scattering. Instead of averaging the scattered intensity over time, the time-correlation function is calculated:

$$C(q,\tau) = \frac{\langle I(q,t+\tau)I(q,t)\rangle}{\langle I^2(q)\rangle} \tag{5.72}$$

where the brackets denote an average over laboratory time. At large values of the time difference, the intensities are uncorrelated and the correlation falls to 1. The measured intensity time-correlation function can be related to the relaxation function for concentration fluctuations:

$$C(q,\tau) = 1 + A\phi_c^2(q,\tau) \tag{5.73}$$

where A is determined by the experimental conditions and by the fraction of the total scattered light due to concentration fluctuations, and

$$\phi_c\left(q,\tau\right) = \frac{\left\langle \Delta c\left(q,\tau\right)\Delta c\left(q,0\right)\right\rangle}{\left\langle \Delta c^2\left(q\right)\right\rangle} . \tag{5.74}$$

The concentration fluctuations in space can be represented in terms of their spatial Fourier components:

$$\Delta c\left(q,\tau\right) = \int_{\text{all }\vec{r}} \left(c\left(\vec{r},\tau\right) - \left\langle c\right\rangle\right)\exp\left(i\vec{q}\cdot\vec{r}\right)d\vec{r} . \tag{5.75}$$

Only concentration fluctuations with the correct spatial component scatter light at any angle (q). Macroscopic diffusion measurements correspond to concentration fluctuations with very low q. The temporal consequences of the large distances over which concentration varies in a cooperative way have restricted the number of such measurements. They take a very long time!

The concentration correlation function can be calculated from Fick's Law. When Equation 5.64 is inserted into Equation 5.63 and the mutual-diffusion coefficient is treated as independent of concentration over the small range of the fluctuations, Fick's Second Law is obtained:

$$\left(\frac{\partial c\left(\vec{r},t\right)}{\partial t}\right) = D_m\nabla^2 c\left(\vec{r},t\right) . \tag{5.76}$$

The quantity of interest requires the spatial Fourier transform:

$$d\ln c\left(q,t\right) = -D_m q^2 dt . \tag{5.77}$$

Integration over the interval τ and multiplication by the concentration fluctuation at zero time, followed by averaging over many fluctuations, yields:

$$\left\langle \Delta c\left(q,\tau\right)\Delta c\left(q,0\right)\right\rangle = \left\langle \Delta c^2\left(q\right)\right\rangle\exp\left(-D_m q^2\tau\right) . \tag{5.78}$$

The concentration fluctuation-relaxation function depends only on the mutual-diffusion coefficient and the square of the scattering vector. Because larger values of q can be obtained by light scattering, the rate of decay of concentration fluctuations can be increased to the kilohertz range. Even when the typical averaging times of 1 min are included, the measurement of the mutual-diffusion coefficient becomes a routine rapid procedure! To ensure the observation of mutual diffusion, the decay constant is measured as a function of q, and the q^2 dependence is verified. The mutual-diffusion coefficient is observed to be a constant with q in dilute solution as long as the value is small enough that $qL_c \ll 1$ and $qR_G < 1$.

When $qR_G > 1$, concentration fluctuations within single macromolecules dominate the correlation function. The calculation of the light-scattering correlation function $S(q,\tau)$ for a single macromolecule in solution is described in detail in Berne and Pecora.[3] The chain is modeled as shown in Equation 5.10. Intramolecular concentration fluctuations resolve into a set of modes with characteristic relaxation times that depend on the subchain friction coefficient, the overall radius of gyration, and the temperature. The correlation function can be expressed as:

$$S(q,\tau) = S_{cm}(q,\tau) S_{intra}(q,\tau) = \exp\left(-D_0 q^2 \tau\right) S_{intra}(q,\tau) ; \tag{5.79}$$

where:

$$S_{intra}(q,\tau) = S(q,\infty) + \left[S(q,0) - S(q,\infty)\right] F(q,\tau) \tag{5.80}$$

$$F(q,\tau) = \sum_{k=1}^{m} A_k(q) \exp\left(-\Gamma_k \tau\right) . \tag{5.81}$$

The q-dependent amplitudes are dominated by the first term for the lowest angles at which internal modes are visible. The values of the relaxation rates have been calculated by Akcasu et al.[1] Observation of the internal modes of random-coil macromolecules in dilute solution by light scattering is now routine. For example, see Kim et al.[23]

5.12 The viscosity of dilute polymer solutions

A continuous fluid is characterized by a local velocity $\vec{v}(\vec{r},t)$. The velocity field is characterized by the velocity gradient tensor $\vec{\nabla}\vec{v}(\vec{r},t)$. The local stress in a fluid due to a local velocity gradient is proportional to the shear viscosity:

$$\sigma_{xy} = \eta\left(\frac{\partial v_x}{\partial y}\right). \tag{5.82}$$

For a Newtonian fluid, this linear relationship is obeyed over a wide range of shear rates (velocity gradients).

The viscosity of a dilute polymer solution can be represented by a virial expansion:

$$\eta = \eta_0 \left(1 + [\eta] c + k_H [\eta]^2 c^2 + \cdots \right) \tag{5.83}$$

where η_0 is the viscosity of the pure solvent, $[\eta]$ is the intrinsic viscosity of the solute, and k_H is called the Huggins coefficient. Another useful formula for the intrinsic viscosity is:

$$[\eta] = \lim_{c \to 0} \frac{\eta - \eta_0}{\eta_0 c}. \tag{5.84}$$

As noted in Chapter 1, the intrinsic viscosity for a macromolecular solute can often be expressed as a power law in the molecular weight:

$$[\eta] = KM^a. \tag{5.85}$$

Because the exponent a is observed to vary over a wide range from 0.0 to 2.0, a theory that explains this behavior would be helpful. The Kirkwood-Riseman treatment given in Section 5.10 is the basis of a successful understanding of the intrinsic viscosity of Gaussian coils in dilute solution.

The first successful theory of the intrinsic viscosity is due to Einstein and was discussed in Chapter 1. It applies to rigid spherical particles in very dilute solution. The intrinsic viscosity is proportional to R^3/M, where R is the radius of the sphere. If the same mass is distributed over a long cylindrical rod, the intrinsic viscosity increases dramatically. The value of $[\eta]$ scales as L^3/M for a rodlike particle.

Why does the intrinsic viscosity depend so dramatically on the size of the particle? The essential microscopic feature of the system that determines the intrinsic viscosity is the ability of the particle to transport momentum from regions of higher velocity to regions of lower velocity. The larger the object, the farther the momentum is transported as the particle rotates in solution.

Consider a long-chain macromolecule composed of m subchains. The location of the center of mass of each subchain is $\vec{R}_j$. The force exerted on the center of mass of each subchain by the fluid was discussed in Section 5.10 for a particular velocity field and a particular polymer conformation (Equation 5.55). The intrinsic viscosity can then be expressed in terms of a sum over all subchains of the x-component of the force exerted on segment j times the y-component of the location of segment j relative to the center of rotation:

$$[\eta] = - \left(\frac{N_A}{M \eta_0 (\partial v_x / \partial y)} \right) \sum_{j=1}^{m} \langle F_{jx} y_j \rangle; \tag{5.86}$$

where the brackets denote an average over all conformations of the chain.

To calculate the intrinsic viscosity, a particular velocity gradient must be selected. Because the relevant quantity is the difference in the local velocity of the fluid and the subchain, it is convenient to note that, in simple shear, the center of mass should be moving at the macroscopic fluid velocity. The local unperturbed relative fluid velocity is then $(\partial v_x / \partial y)y$. Using the same procedures introduced for the friction coefficient, the required average can be expressed as:

$$\langle F_{jx}y_j \rangle = -\left(\frac{\zeta}{2}\right)\left(\frac{\partial v_x}{\partial y}\right)\langle y_j y_j \rangle - \left(\frac{\zeta}{6\pi\eta_0}\right)\sum_{k\neq j}^{m}\left\langle \frac{1}{R_{jk}}\right\rangle \langle F_{kx}y_j \rangle. \tag{5.87}$$

The solution to this set of coupled equations proceeds along the same lines as the calculation of the friction coefficient. The intrinsic viscosity can be expressed as:

$$[\eta] = \left(\frac{N_A \zeta m^2 \langle R_{12}^2 \rangle}{36\eta_0 M}\right)F(X) \tag{5.88}$$

where X is as defined in Equation 5.60 and the function of X is:

$$F(X) = \left(\frac{6}{\pi^2}\right)\sum_{k=1}^{\infty}\frac{1}{k^2\left(1+\dfrac{X}{k^{1/2}}\right)}. \tag{5.89}$$

Within the free-draining ($X = 0$) limit, the intrinsic viscosity can be written as:

$$[\eta](X=0) = \left(\frac{N_A \zeta m}{6\eta_0 M}\right)\langle R_G^2 \rangle_0 \tag{5.90}$$

for Gaussian unperturbed chains. This limit is not reasonable for long chains in dilute solution. In the nondraining limit:

$$[\eta] = 6^{3/2}\Phi_0\left(\frac{\langle R_G^2 \rangle_0^{3/2}}{M}\right) \tag{5.91}$$

where the numerical constant has the value $\Phi_0 = 2.87 \times 10^{23}$. For Gaussian chains within the nondraining limit, the intrinsic viscosity scales as $M^{1/2}$.

This is the correct result observed in the laboratory under conditions where the osmotic second virial coefficient vanishes.

It has also been found useful to write the intrinsic viscosity in the form:

$$[\eta] = \left(\frac{K\pi N_A}{M}\right)\langle R_G^2 \rangle R_H \qquad (5.92)$$

where the constant K depends on the draining conditions and the hydrodynamic shape of the particle. For an expanded coil in a good solvent, the mean-squared radius of gyration scales as $M^{6/5}$. The hydrodynamic radius under the same conditions might increase with a power law anywhere from $M^{1/2}$ to M. The observed molecular-weight dependence of the intrinsic viscosity in a good solvent has an exponent within the range 0.7 to 1.0. The full understanding of the intrinsic viscosity is still elusive, but reasonable results can be obtained by treating the random-coil macromolecule as a spherical particle with the radius proportional to the radius of gyration. The Huggins coefficient is due to the intermolecular hydrodynamic interactions. The theory is exceedingly difficult, but experimental determinations of k_H yield values in the range 0.3 to 0.7.

5.13   The effect of molecular-weight polydispersity

The value of the molecular quantities discussed in Chapter 5 all depend on the molecular weight of the polymer. Many actual samples of polymeric material are characterized by a distribution of molecular weights. In this section, the effect of polydispersity in the molecular weight will be discussed for each of the physical properties introduced above.

Consider a sample of polymeric material with the same composition but with a distribution of chain lengths. Let there be N molecules in the sample. Each molecule is characterized by a mass, m_i. The molecular weight for each species is given by $M_i = m_i N_A$. When all the molecules are weighed, they are grouped by molecular weight and enumerated as $\{N_i\}$. The molecular-weight distribution is defined by the mole fraction of each species:

$$\wp(M_i) = \frac{N_i}{N} = x_i . \qquad (5.93)$$

The average molecular weight for the sample is defined as:

$$\langle M \rangle = \sum_{\text{all } i} \wp(M_i) M_i . \qquad (5.94)$$

Because the probability distribution is based on the number fraction, the average is called the number average molecular weight. It is the normal mathematical definition of an average.

Many properties depend on the weight fraction instead of the mole fraction. The weight fraction can be defined as:

$$w_i = \frac{N_i M_i}{\sum_{\text{all } i} N_i M_i} . \tag{5.95}$$

The weight-averaged molecular weight is then defined as:

$$\left\langle M \right\rangle_w = \sum_{\text{all } i} w_i M_i . \tag{5.96}$$

In a solution, the weight fraction of the solute particles is expressed directly in terms of the concentration: $w_i = c_i/c$, where c_i is the concentration of molecular-weight species, i. The weight-averaged molecular weight can then be expressed as:

$$\left\langle M \right\rangle_w = \left(\frac{1}{c} \right) \sum_{\text{all } i} c_i M_i . \tag{5.97}$$

There can be as many weighting functions are there are physical properties. We will consider each measurable property and analyze the effect of a distribution of molecular weights on the interpretation of results.

Consider the measurement of the excess Rayleigh scattering in a dilute polymer solution. The expression for the scattering of a group of independent particles within the limit of infinite dilution is:

$$\Delta R\left(q, \{c_i\}\right) = K \sum_{\text{all } i} c_i M_i S_i \left(q\right) \tag{5.98}$$

where K is the usual optical constant (Equation 5.30). Within the thermodynamic ($q = 0$) limit, the excess Rayleigh ratio can be expressed as:

$$\Delta R\left(q = 0, \{c_i\}\right) = Kc \left\langle M \right\rangle_w \tag{5.99}$$

by comparing with Equation 5.97 and observing that $S(0) = 1$. The weight-averaged molecular weight is obtained by light scattering within the limit of infinite dilution and $q = 0$. At low q, the single-molecule scattering function is given by Equation 3.9. The excess scattering function can then be expressed as:

$$\Delta R\left(q \to 0, \{c_i\}\right) = Kc\langle M\rangle_w - \left(\frac{Kq^2}{3}\right)\sum_{\text{all } i} c_i M_i \langle R_G^2\rangle_i. \tag{5.100}$$

The radius of gyration is weighted by both the concentration and the molecular weight. The z-fraction weighting function is defined as:

$$z_i = \frac{N_i M_i^2}{\displaystyle\sum_{\text{all } i} N_i M_i^2} = \frac{c_i M_i}{\displaystyle\sum_{\text{all } i} c_i M_i}. \tag{5.101}$$

The overall excess scattering function can then be expressed as:

$$\Delta R\left(q \to 0, \{c_i\}\right) = Kc\langle M\rangle_w \left(1 - \left(\frac{q^2 \langle R_G^2\rangle_z}{3}\right)\right) \tag{5.102}$$

where

$$\langle R_G^2\rangle_z = \left(\frac{1}{c\langle M\rangle_w}\right)\sum_{\text{all } i} c_i M_i \langle R_G^2\rangle_i; \tag{5.103}$$

is the z-averaged mean-squared radius of gyration!

Within the thermodynamic limit in dilute solution, the excess Rayleigh ratio can be expressed as:

$$\Delta R\left(q = 0, c \to 0\right) = Kc\langle M\rangle_w - 2K\sum_{\text{all } i} c_i^2 M_i^2 A_{2i}. \tag{5.104}$$

While a weighting function can be defined for this sum, it is best to simply realize that the second virial coefficient for a polydisperse sample is strongly weighted in favor of the highest molecular weights.

The osmotic pressure also depends on the molecular weight of the solute. The van't Hoff Law for a mixture of solute species can be written as:

$$\pi\left(T,\{c_i\}\right) = RT \sum_{\text{all } i} \frac{c_i}{M_i} \,. \tag{5.105}$$

The mole fraction can be written in terms of the concentration as:

$$x_i = \frac{\dfrac{c_i}{M_i}}{\displaystyle\sum_{\text{all } i} \dfrac{c_i}{M_i}} \,. \tag{5.106}$$

The number average molecular weight is then:

$$\langle M \rangle = \frac{c}{\displaystyle\sum_{\text{all } i} \dfrac{c_i}{M_i}} \,. \tag{5.107}$$

The van't Hoff Law then becomes:

$$\pi\left(T,\{c_i\}\right) = \frac{cRT}{\langle M \rangle} \,. \tag{5.108}$$

The limiting osmotic pressure depends on the number average molecular weight. At finite concentration the osmotic pressure becomes:

$$\pi\left(T,\{c_i\} \to 0\right) = \frac{cRT}{\langle M \rangle} + RT \sum_{i,j} c_i c_j A_{2ij} \tag{5.109}$$

where A_{2ij} is the mixed second virial coefficient for different molecular weights. The average of the second virial coefficient obtained from the analysis of osmotic pressure data is different from the average obtained from the analysis of light-scattering data!

The self-diffusion coefficient, D_s (Equation 5.48), depends on the molecular weight through the chain-length dependence of the hydrodynamic radius R_H. A common empirical expression for this quantity is

$$R_H = KM^a \tag{5.110}$$

If dynamic light scattering is used to measure the diffusion coefficient, the expression for the scattering function that determines the molecular-weight average is:

$$S(q,t) = \sum_{\text{all } i} z_i \exp(-D_i q^2 t).$$

(5.111)

While the observed decay function could be analyzed with Equation 5.111, prior knowledge of the shape of the distribution would be required to determine a set of $\{z_i\}$. If the molecular-weight distribution is narrow, the decay function can be analyzed by the method of cumulants. The logarithm of the decay function, within the limit of small values of the argument, can be expressed as:

$$\ln S(q,t) =$$

$$-\sum_{\text{all } i} z_i \frac{k_b T q^2 t}{6\pi\eta R_{H,i}} + (1/2)\sum_{\text{all } i} z_i D_i^2 q^4 t^2 - (1/2)\left[\sum_{\text{all } i} z_i D_i q^2 t\right]^2 + \cdots$$

(5.112)

While this is a transparent expression for the results of a light-scattering experiment, it is clear that the molecular-weight average obtained by this technique is not simple. The z-average diffusion coefficient is obtained from the first term. However, the average of the hydrodynamic radius is the z-average inverse hydrodynamic radius. The higher-order terms in time yield the z-averaged variance of the diffusion distribution.

The relative excess viscosity of a polymer solution can be expressed in the form:

$$\frac{\eta - \eta_0}{\eta_0} = \sum_{\text{all } i} c_i [\eta]_i = c \langle [\eta] \rangle_w.$$

(5.113)

The weight-averaged intrinsic viscosity is obtained. If the molecular-weight dependence of the intrinsic viscosity of monodisperse chains (Equation 5.85) is inserted into Equation 5.113, the observed weight-averaged intrinsic viscosity can be expressed as:

$$\langle [\eta] \rangle_w = K \sum_{\text{all } i} w_i M_i^a = K \langle M \rangle_v^a$$

(5.114)

where the new average is called the viscosity-averaged molecular weight. Because intrinsic viscosity is relatively easy to measure, the viscosity-averaged molecular weight has gained a firm position in the polymer literature, but it is not a simple quantity!

If the distribution of molecular weights is narrow, the best approach to measuring the magnitude of the variance in the distribution is dynamic light scattering. However, this approach is time consuming and requires a detailed knowledge of the relationship between the diffusion coefficient and the molecular weight of the solute particles. The technique that is routinely used to measure $\wp(M)$ when the distribution is sufficiently broad is size-exclusion chromatography (SEC). This approach allows the solute particles to be separated on the basis of their hydrodynamic volumes:

$$V_H = K\left[\eta\right]M \tag{5.115}$$

where K is a constant for each type of particle. Attempts to establish a universal value for K have not been entirely successful. A dilute solution of particles is forced to flow through a heterogeneous medium containing large particles with small pores. Smaller solute particles spend a longer time exploring the pores than larger ones due to entropic exclusion. The elution volume of each species is monitored with a detector that is sensitive to the presence of solute.

The most common detector is a differential refractometer, which then is proportional to the concentration of polymer at that elution volume. More-sophisticated detectors include a low-angle, light-scattering device that, in tandem with the differential refractometer, yields a direct measure of molecular weight associated with particles at the observed elution volume. Size-exclusion chromatographs are now highly engineered instruments that have proven invaluable to research on macromolecules.

chapter six

Structure and properties of polymers in semidilute solution

6.1 Introduction

Long-chain molecules in dilute solution influence a region of space large with respect to the sum of the volumes of the individual mers. This concept was expressed in terms of the pervaded volume, $V_P = (4\pi/3)\langle R_G^2 \rangle^{3/2}$, in Chapter 5. When the sum of the pervaded volumes of the solute macromolecules equals the macroscopic solution volume, the solution can no longer be considered dilute. It is convenient to define a crossover concentration (c^*) to the semidilute regime as:

$$c^* = \frac{M}{N_A V_P}.$$
(6.1)

It will be shown that many of the properties of polymer solutions in the semidilute regime depend in a detailed way on the value of the crossover concentration, c^*.

Two polymer molecules resist overlapping because the presence of one chain restricts the conformations available to the other chain. This entropically based intermolecular repulsion exists in addition to the segment-segment excluded volume introduced in Chapter 5. If the chains are dissolved in a "good" solvent, the mean-squared radius of gyration is increased in dilute solution due to the intramolecular excluded volume, as discussed in Section 5.2. As the concentration is increased into the semidilute regime, intermolecular repulsion must occur, and the chains are observed to contract in order to reduce intermolecular repulsive energy. It is observed experimentally that the mean-squared radius of gyration in semidilute solutions under good solvent conditions can be described by a scaling law:

$$\left\langle R_G^2 \right\rangle \sim c^{-1/4} \,. \tag{6.2}$$

This phenomenon is explained in detail in Section 6.3. Many properties of polymer solutions in the semidilute regime are observed to be described by power-law relationships, and the scaling symbol, $\sim$, reflects this mathematical fact.

The van't Hoff Law for osmotic pressure (Equation 5.19) depends explicitly on the molecular weight of the solute. One of the most remarkable properties of semidilute solutions is that the osmotic pressure is observed to be independent of the molecular weight of the macromolecules. Another property that is observed to be independent of molecular weight in semidilute solution is the mutual-diffusion coefficient, $D_m(c)$. These phenomena are discussed in Section 6.2 and explained in more detail in Section 6.3.

The universal behavior of many properties in semidilute solution has been explained by deGennes.[9] He proposed that there is a universal length in this concentration range that depends only on the polymer concentration and the nature of the polymer–solvent interactions. The deGennes theory of semidilute solutions is presented in Section 6.3 and used to explain many observed features of these systems.

The success of the deGennes theory suggests that the viscosity in semidilute solutions can also be explained in terms of the universal length and the actual radius of gyration of the macromolecules. This theory is presented in Section 6.4.

Polymer solutions also display many interesting features in the crossover regimes from dilute to semidilute and from semidilute to concentrated solutions. The observations and the concepts necessary to comprehend these phenomena are discussed in Section 6.5.

6.2 The remarkable behavior of semidilute solutions

The osmotic pressure of a semidilute polymer solution can be described by a power law in the concentration. It is convenient to express this relationship as:

$$\pi\left(T,c\right) \sim \left(\frac{cRT}{M}\right)\left(\frac{c}{c^*}\right)^x \tag{6.3}$$

where the exponent is observed to depend on the solvent quality. The value of the exponent can be inferred from the observation that the osmotic pressure is independent of the molecular weight. For a good solvent, the crossover concentration scales as:

$$c^* \sim \frac{M}{\langle R_G^2 \rangle^{3/2}} \sim M^{1-3v} \sim M^{-4/5} \,. \tag{6.4}$$

This implies that, in a good solvent, the osmotic pressure in the semidilute regime scales as:

$$\pi(T,c) \sim c^{9/4} \tag{6.5}$$

which is exactly what is observed experimentally. At the Flory temperature, the crossover concentration scales as $M^{-1/2}$, which implies that the osmotic pressure scales as c^3, which is also what is observed experimentally. The success of the scaling theory in explaining the concentration dependence of the osmotic pressure in the semidilute regime suggests that it may be useful for other properties.

The mutual-diffusion coefficient in semidilute polymer solutions can be expressed as:

$$D_m(T,c) \sim \left(\frac{k_b T}{6\pi\eta_0 R_H} \right) \left(\frac{c}{c^*} \right)^x \tag{6.6}$$

where the pure solvent viscosity is retained. This relationship emphasizes that the semidilute regime is restricted to small-volume fractions for the polymer. It is only the large size of the pervaded volume that yields overlap at small values of the volume fraction of polymer. Within the good-solvent limit, and the nonfree-draining limit, the hydrodynamic radius scales as M^v, which implies that:

$$D_m(c) \sim c^{3/4} \,. \tag{6.7}$$

The predicted scaling law is approximately obeyed in good solvents. Similar considerations at the Flory temperature predict that the mutual-diffusion coefficient scales as c, which is a good representation of the measured data. While the scaling theory is not quite as accurate for a transport property as it was for a thermodynamic property, it is still an excellent approximation. The observation of universal scaling behavior for polymer solutions in the semidilute concentration regime suggests that there must be some corresponding universal quantity associated with these systems.

6.3 *Microscopic theory of semidilute solutions*

To explain the behavior of polymer solutions in the overlapped regime, it is necessary to adopt a picture of the structure of these systems. The chain molecule is represented as a string of m statistical subunits with mean-squared end-to-end vectors $\langle R_{12}^2 \rangle$. The average concentration inside a statistical subunit can be expressed as:

$$c_s = \frac{M}{mN_A (\pi/6)\langle R_{12}^2 \rangle^{3/2}} . \tag{6.8}$$

One arbitrary definition of the semidilute regime is the concentration range: $c^* < c < c_s$.

As the chains start to overlap, the individual subunits still tend to avoid one another. A useful quantity introduced by deGennes[9] is the intramolecular pair-correlation function for subunits inside a single chain. Consider a particular subunit at the origin surrounded by a sphere of radius r. The concentration of polymer inside the sphere can be calculated if we can determine the number of subunits that, on average, will be found within the sphere. For a good solvent, the mean-squared end-to-end distance of a chain, here taken to be r^2, scales as $n^{2\nu} \langle R_{12}^2 \rangle$. This means that the average concentration within a sphere of radius r scales as:

$$g(r) \sim \frac{M}{mN_A (4\pi/3)\langle R_{12}^2 \rangle^{5/6} r^{4/3}} . \tag{6.9}$$

As long as the intramolecular concentration is larger than the average solution concentration, the correlation function will decrease as shown in Equation 6.9. When the average solution concentration exceeds the overlap concentration, there will be a radius such that subunits separated by more than some limiting number of units are just as likely to be near subunits on a different chain. The entropic repulsion of subunits on the same chain is then screened by the presence of subunits on other chains. This analysis leads to the concept of a screening length, ξ, for the excluded-volume effect in a semidilute solution. Inside regions of volume $(4\pi/3)\xi^3$, all the subunits are likely to be from the same chain. In a good solvent, the number of subunits inside such a region scales as:

$$n \sim \frac{\xi^{5/3}}{\langle R_{12}^2 \rangle^{5/6}} . \tag{6.10}$$

There is no longer a tendency for subunits in different regions to be repelled, and it is convenient to view the overall chain as a sequence of subregions of mean-squared end-to-end length ξ^2 that can be described by random-coil statistics. The number of such subregions should scale as m/n. The mean-squared end-to-end length of the chain should now scale as:

$$\langle R^2 \rangle \sim (m / n)\xi^2 \sim m\langle R_{12}^2 \rangle^{5/6} \xi^{1/3} . \tag{6.11}$$

As the screening length decreases in a good solvent, the chain dimensions should also decrease until the screening length is comparable to the statistical subunit length.

To satisfy the criterion of universality, the screening length should depend only on the concentration of the solution and be independent of molecular weight. The screening length can be expressed as a scaling relationship:

$$\xi \sim \langle R_G^2 \rangle^{1/2} \left(\frac{c}{c^*} \right)^x . \tag{6.12}$$

In a good solvent, the screening length scales as $c^{-(3/4)}$, which leads to a concentration dependence for the mean-squared end-to-end length, $c^{-(1/4)}$, which is consistent with experimental observations.

Another important quantity that depends explicitly on the screening length is the concentration-fluctuation correlation function:

$$g_c(r) = \left(\frac{1}{\langle c \rangle} \right) \left[\langle c(0)c(r) \rangle - \langle c \rangle^2 \right] . \tag{6.13}$$

For values of r less than the screening length, the correlation function should scale as in Equation 6.9, since only intrachain correlations are involved. For distances greater than the screening length, deGennes has proposed that the concentration-fluctuation correlation function scales as:

$$g_c(r) \sim \langle c \rangle \left(\frac{\xi}{r} \right) \exp\left(-\frac{r}{\xi} \right) . \tag{6.14}$$

The spatial Fourier transform of the concentration correlation function is the light-scattering structure factor $S(q,c)$. The predicted form is then:

$$S(q,c) \sim \frac{\langle c \rangle \xi^3}{1 + q^2 \xi^2} . \tag{6.15}$$

Neutron-scattering measurements on polymer solutions in the semidilute regime are well described by Equation 6.15 for small values of the product $q\xi$, and the measured screening length is observed to decrease as $c^{-(3/4)}$ in good solvents. The success in predicting the concentration dependence of the chain dimensions and the direct observation of a length that behaves like the predicted screening length suggests that the concept introduced by deGennes is useful in the description of the universal behavior in semidilute solutions.

Chains must also overlap in solutions at the Flory temperature. Application of Equation 6.12 to this case predicts that the screening length should decrease as $1/c$ for these solutions. The overall chain dimensions are observed to be independent of concentration for solutions at the Flory temperature. This independence is also predicted if the above concentration dependence of the screening length is assumed, using Equation 6.11, and the number of subunits in a subregion is calculated based on random-coil statistics:

$$n_\theta \sim \frac{\xi^2}{\langle R_{12}^2 \rangle} . \tag{6.16}$$

Application of the model of a semidilute solution as a collection of regions of size ξ can be made to both the osmotic pressure and the mutual-diffusion coefficient. It has been proposed by deGennes[9] that the osmotic pressure should scale as the number density of screening regions:

$$\pi(T,c) \sim \frac{T}{\xi^3} \sim Tc^{9/4}(\text{good}) \sim Tc^3(\theta) ; \tag{6.20}$$

which correctly predicts the observed concentration dependencies. The regions are screened thermodynamically from one another. It has also been proposed that the mutual-diffusion coefficient should be described as

$$D_m(c) = \frac{c\left(\dfrac{\partial \pi}{\partial c}\right)}{\Xi} ; \tag{6.21}$$

where Ξ is the mutual friction per unit volume. The mutual friction per unit volume is calculated as the number of regions per unit volume times the friction per region. The friction per region is proposed to scale as the size of the region. If a good solvent is considered, the mutual-diffusion coefficient should scale as:

$$D_m\left(\text{good}\right) \sim \frac{c^{9/4}}{c^{3/2}} \sim c^{3/4} \;. \tag{6.22}$$

Similar considerations at the Flory temperature yield:

$$D_m\left(\theta\right) \sim \frac{c^3}{c^2} \sim c \;; \tag{6.23}$$

which is also what is observed experimentally. Whereas the mutual-diffusion coefficient decreases with concentration in dilute solution at the Flory temperature, it increases in semidilute solution. The concept of thermodynamic and hydrodynamic screening allows a rational discussion of the properties of semidilute polymer solutions.

6.4   Viscosity in semidilute solutions

While the regions are thermodynamically and hydrodynamically screened, the actual chains connect regions and influence the viscosity. It has been proposed that the viscosity in the semidilute regime can be described as:

$$\eta\left(c\right) \sim \eta_0 c \left(\frac{\left\langle R_G^2 \right\rangle}{M}\right) R_H \;. \tag{6.24}$$

This scaling relation requires that the solution still be quite dilute so that the solvent viscosity remains the dominant local variable and that the chains have not overlapped enough to display evidence of entanglements.

At the Flory temperature, the radius of gyration is proportional to M and is independent of concentration. The hydrodynamic radius is proportional to the screening length times the number of subregions per molecule, m/n_θ. The viscosity under these conditions is predicted to scale as:

$$\eta_\theta \sim \eta_0 M c^2 \;. \tag{6.25}$$

Under good solvent conditions, both the radius of gyration and the hydrodynamic radius are functions of concentration. Application of Equations 6.10 and 6.11 leads to a prediction:

$$\eta\left(\text{good}\right) \sim \eta_0 M c^{5/4} \;. \tag{6.26}$$

Because the chains are shrinking as the concentration is increased, the viscosity increases less strongly with c.

6.5 Structure near overlap

The structure of the solution near the overlap concentration can be studied using light scattering. As the concentration increases, the scattering from different molecules interferes and the value of $S(0)$ decreases. For good solvents, the molecules behave like soft repulsive spheres. As the concentration approaches c^*, the solution of soft spheres is similar to that of a liquid. Consider the scattering functions shown in Figure 6.1. The overlap concentration for this system is 2.9 mg/ml. As the concentration approaches c^*, the scattering function displays a maximum with angle. This is indicative of liquidlike structure in the solution. As the solution becomes even more overlapped, the absolute values of $S(q,c)$ become even lower, and the maximum is no longer obvious. The structure of the solution crosses over from a dilute solution, where the chains in a good solvent are expanded by the excluded-volume interaction, to a semidilute solution, where there is no thermodynamic trace of the size of individual chains. Near the overlap concentration, the size of the chains is still evident, and the structure of the solution is well described by a liquidlike pair-correlation function for the molecular centers of mass.[20]

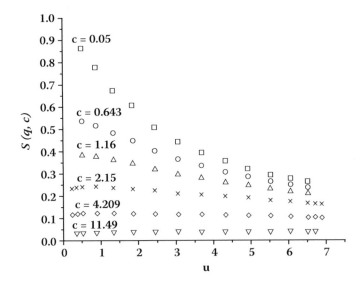

Figure 6.1 Scattering functions, $S(q,c)$, of poly(α-methyl styrene) in toluene as a function of concentration near overlap. The dimensionless abscissa is $u = q^2 \langle R_G^2 \rangle$.

chapter seven

Structure and properties of polymers in concentrated solution

7.1 Introduction

In a dilute solution, the pervaded volume of the molecule is small with respect to the available solution volume per molecule: M/cN_A. In a semidilute solution, the volume fraction of the polymer remains small, but the pervaded volume exceeds the solution volume per molecule. In a concentrated solution, the volume fraction of the polymer solute is large enough that, on average, each solvent molecule has at least one mer as a nearest neighbor. The concept of a semidilute solution is predicated on a very long chain where the number of mers is large with respect to the length at which the characteristic ratio reaches its limiting value. Under these conditions, the chain can be divided into a large number of statistical subchains. When the screening length of the solution falls to a value comparable with the root mean squared (rms) radius of gyration of such a limiting subchain, the structure of the chain is dominated by intramolecular factors, and the solvent quality plays only a local role in the conformation of the polymer. This is the concentrated regime. Since the screening length is now a very small fraction of the rms radius of gyration, the local structure of the polymer solution is essentially independent of molecular weight. On length scales large with respect to the screening length, no information about the structure of the solution can be obtained from thermodynamic or static light-scattering measurements.

The mean-squared radius of gyration of a linear chain molecule in the concentrated-solution regime is proportional to the number of mers in the macromolecule. The thermodynamic properties of the solution depend on the interaction between the solvent molecules and the mers. To facilitate the discussion of these interactions, Flory chose to divide the chain molecule into subunits based on the ratio of the partial volumes of the polymer and the solvent: $x = v_2/v_1$. The total volume of the solution composed of N_1 molecules of solvent and N_2 molecules of polymer is:

$$V = N_1 v_1 + N_2 v_2 . \tag{7.1}$$

The solution is also characterized by the number of nearest neighbors for each solvent molecule, z. This simple lattice model is able to predict many of the thermodynamic properties of concentrated polymer solutions. For the local structure to be in the so-called mean field limit, the polymer volume fraction should exceed approximately 10%.

The details of the Flory–Huggins theory of polymer solutions are presented in Section 7.2. While quantitative agreement with experimental results is only fair, the theory does predict an upper critical solution temperature for solutions in poor solvents. Explicit calculations of the spinodal curve and the binodal coexistence curve are presented.

The Flory–Huggins theory can also be used to describe the thermodynamics of gels. The Gibbs energy of mixing for the gel is approximated by the ideal-rubber elastic theory presented in Chapter 4 plus the Gibbs energy of mixing presented in this chapter. The Flory–Rehner equation for swelling equilibrium is derived and discussed in Section 7.3.

The thermodynamic properties of concentrated solutions can be measured using light scattering. The application of the Flory–Huggins theory to this experimental situation is presented in Section 7.4

Experimental studies of concentrated polymer solutions reveal that most binary liquid mixtures display a lower critical solution temperature. A theory including equation of state contributions was developed by Orwoll and Flory.[31] This theory is presented and discussed in Section 7.5

High-molecular-weight chains in concentrated polymer solutions diffuse at a very slow rate. A theory developed by deGennes to calculate the self-diffusion coefficient of highly entangled chains is presented in Section 7.6. The mutual-diffusion coefficient for concentrated polymer solutions is also derived and discussed.

The combination of high molecular weight and high concentration leads to a very dramatic increase in the viscosity of the solution. A theory elaborated by deGennes[9] is presented in Section 7.7. In addition, for rapid deformation processes, the concentrated solution displays viscoelastic behavior. One of the most challenging problems in polymer science is the full explanation of the viscoelastic behavior of high-molecular-weight concentrated solutions.

7.2   Flory-Huggins theory of concentrated solutions

The phase behavior of a two-component solution depends on the chemical potentials of the solvent and the solute. While it would be very difficult to calculate the chemical potential of the pure solvent, the osmotic pressure depends only on the difference in the chemical potential due to the mixing with the solute (Equation 5.12). Flory and Huggins independently developed an approximate theory for the Gibbs energy of mixing of two liquids. It is important to note that it is assumed that the intramolecular entropy of the solvent and the polymer solute are independent of the mixing process. It is also assumed that the volume change upon mixing vanishes. Finally, the mixing is assumed to be completely random. In spite of these approximations, and many more of a more subtle nature, the theory gives qualitative insight into the nature of these solutions.

The Gibbs energy of mixing, $\Delta G_{mix}(T,N_1,N_2)$, can be divided into an entropy of mixing and an enthalpy of mixing. For a random mixture, the entropy of mixing was shown by Flory[15] to be given by:

$$\Delta S_{mix}\left(N_1, N_2\right) = -k_b \left[N_1 \ln \phi_1 + N_2 \ln \phi_2\right] \tag{7.2}$$

where ϕ_2 is the volume fraction of the solute. The realization that volume fractions were the key measures of composition already explained many of the properties of polymer solutions. The observation of phase separation in polymer solutions requires that the enthalpy of mixing be considered because the ideal entropy of mixing leads to miscibility in all proportions for two pure liquids. The theory is developed in terms of a local interaction enthalpy for nearest neighbors. If a solvent molecule interacts with another solvent molecule, an enthalpy w_{11} is assigned to the interaction. The interaction enthalpy must be negative for two neutral molecules on neighboring sites in the solution. If the neighboring unit is one of the x subchains of the polymer, an enthalpy w_{12} is assigned. The corresponding mer–mer interaction is w_{22}. The enthalpy of mixing can be expressed in terms of the exchange enthalpy:

$$\Delta w_{12} = w_{12} - \frac{w_{11} + w_{22}}{2} . \tag{7.3}$$

The exchange enthalpy can be either positive or negative, but it is usually positive because like interactions are more favorable than unlike interactions. The total enthalpy of mixing can be calculated for the N_1 solvent molecules by considering the averaged behavior of one solvent molecule. The number of nearest neighbors that are solute mers is given by the total number of nearest neighbors, z, times the volume fraction of polymer. For each such neighbor, the enthalpy of mixing is incremented by the exchange enthalpy. The total enthalpy of mixing is then:

$$\Delta H_{mix} = N_1 z \phi_2 \Delta w_{12} . \tag{7.4}$$

Since the enthalpy of mixing contains two unknown parameters that always appear as a product, it is convenient to define a quantity that will appear constantly in the theory anyway:

$$\chi = \frac{z\Delta w_{12}}{k_b T} . \tag{7.5}$$

The Flory χ parameter has found widespread use in the discussion of any two-component mixture, including polymer mixtures. The Gibbs energy of mixing can then be written as:

$$\Delta G_{mix} = k_b T \left[N_1 \ln \phi_1 + N_2 \ln \phi_2 + N_1 \phi_2 \chi \right] . \tag{7.6}$$

The Flory–Huggins expression for the Gibbs energy of mixing allows a calculation of the chemical potential of the solvent:

$$\mu_1 - \mu_1^0 = \left(\frac{\partial \Delta G_{mix}}{\partial N_1} \right)_{T,N_2} = k_b T \left[\ln \left(1 - \phi_2 \right) + \phi_2 \left(1 - \frac{1}{x} \right) + \phi_2^2 \chi \right] \tag{7.7}$$

where the relation $\phi_1 = 1 - \phi_2$ has been used. The osmotic pressure is then:

$$\pi(T, \phi_2) = -\left(\frac{k_b T}{v_1} \right) \left[\ln \left(1 - \phi_2 \right) + \phi_2 \left(1 - \frac{1}{x} \right) + \phi_2^2 \chi \right] . \tag{7.8}$$

The osmotic pressure expression will be used to analyze the phase behavior of a polymer solution.

Although the Flory–Huggins theory is not truly valid at low-volume fractions of solute, it is useful to examine the dilute limiting law for the osmotic pressure. The expression $\ln(1 - x) = -x - x^2 / 2 - \cdots$ is used:

$$\pi_{\phi_2 \to 0} = \left(\frac{k_b T}{v_1} \right) \left[\left(\frac{v_1}{v_2} \right) \phi_2 + \left(\frac{1}{2} - \chi \right) \phi_2^2 + \cdots \right] . \tag{7.9}$$

The van't Hoff Law is recovered in the limit of infinite dilution, since $\phi_2 = N_2 v_2 / V$. The second virial coefficient is zero when $\chi = 1/2$. The identification of the Flory temperature as the point where the Flory χ parameter is $1/2$ helps to guide the discussion of the behavior of the osmotic pressure as a function of χ.

When the Flory χ parameter vanishes, the mixing is athermal and the solvent is considered very good! For most solvents, the mixing is endothermic, and there will exist a temperature at which $\chi = 1/2$. For higher temperatures, the solution is predicted to be homogeneous at all compositions. For lower temperatures, the value of χ exceeds 1/2, and the solution may separate into a dilute phase and a more concentrated phase under certain conditions. For low values of χ, the osmotic pressure rises monotonically with the volume fraction of the polymer. For a critical value, χ_c, there exists a composition where $(\partial \pi / \partial \phi_2)_{T,x} = 0$. Under these conditions, there are large fluctuations in composition in the solution. For larger values of χ, there will be two compositions at each temperature where the derivative is zero. The locus of such points is called the spinodal curve for the solution. Concentrations between these two limits are not allowed, since the condition, $(\partial \pi / \partial \phi_2)_{T,x} < 0$, is mechanically unstable. For the Flory–Huggins theory, the spinodal curve is calculated to be:

$$\chi_s\left(\phi_2\right) = -\frac{\left(1-\left(1/x\right)\right)}{2\phi_2} + \frac{1}{2\phi_2\left(1-\phi_2\right)}. \tag{7.10}$$

The shape of this curve depends on the value of x. The minimum value of χ_s can be found by differentiation. The value of ϕ_{2c} obtained by this procedure is:

$$\phi_{2c} = \frac{1}{1+x^{1/2}}. \tag{7.11}$$

Inserting Equation 7.11 into Equation 7.10 yields:

$$\chi_c = \chi_s\left(\phi_{2c}\right) = \frac{\left(1+x^{1/2}\right)^2}{2x}. \tag{7.12}$$

The critical value of χ in the Flory–Huggins theory is a function of x alone. In the limit of large x:

$$\lim_{x \to \infty} \chi_c = \frac{1}{2} + \frac{1}{x^{1/2}}. \tag{7.13}$$

For high-molecular-weight polymers, the critical value of χ approaches the value of χ at the Flory temperature, within the limit of large x. Thus the critical temperature is equal to the Flory temperature.

The critical volume fraction is very small for high-molecular-weight solutes. In fact, the critical volume fraction is comparable to c^*. When the

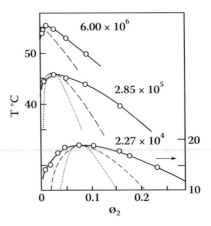

Figure 7.1 Experimental phase diagrams for polyisobutylene fractions in diisobutyl ketone. Also shown are theoretical Flory–Huggins binodals (dashed) and spinodal (dotted) curves. (From Cassasa, E.F., Phase Equilibrium in Polymer Solutions, in *Fractionation of Synthetic Polymers*, Tung, L.H., Ed., Marcell Dekker, New York, 1977. With permission.)

solvent is poor, phase separation starts very close to the Flory temperature for concentrated solutions. The solution then separates into a dilute solution and a more concentrated phase. While the spinodal curve defines the limits of mechanical stability for the solution, the concentrations of the two phases at equilibrium must be determined by the conditions that the chemical potentials of the solvent in the two phases are equal and that the chemical potentials of the polymer in the two phases must be equal. The excess chemical potential of the polymer is given by:

$$\mu_2 - \mu_2^0 = k_b T \left[\ln \phi_2 + \left(1 - \frac{v_2}{v_1}\right)(1 - \phi_2) + \chi \left(\frac{v_2}{v_1}\right)(1 - \phi_2)^2 \right]. \qquad (7.14)$$

The locus of equilibrium volume fractions as a function of χ is called the binodal. The binodal and the spinodal curves merge at the critical point. A plot of the binodal curve for several molecular weights is shown in Figure 7.1. The region between the spinodal and binodal curve is metastable with regard to phase separation. Since the mutual-diffusion coefficient of the solution determines the rate of phase separation, the high viscosity of concentrated polymer solutions will retard this process. A discussion of the dynamics of concentrated polymer solutions is given in a later section. Since the Flory–Huggins theory is explicitly invalid for concentrations in both the dilute and semidilute regimes, the calculated phase diagrams are expected to differ from those obtained experimentally. However, the qualitative shape of these curves is instructive, just as the van der Waals theory of fluids yields insight into the phase behavior of gases and liquids.

7.3 The thermodynamics of swollen rubber: gels

One application of the Flory–Huggins theory of concentrated solutions is to the system composed of a rubber network swollen by solvent: a gel. The chemical potential of the solvent in the gel can be calculated with respect to the pure solvent if the Gibbs energy of the gel can be obtained. The approximation introduced by Flory is that the free energy of swelling can be represented as the free energy of deformation of the network plus the Flory–Huggins free energy of mixing:

$$\Delta G_{swell} = \Delta G_{el} + \Delta G_{mix} \; . \tag{7.15}$$

The elastic Gibbs energy introduced in Chapter 4 must be extended to include the swelling deformation, since the volume of the network will now change. The deformation ratio for isotropic swelling is defined as: $\alpha^3 = V/V_0$, where V_0 is the volume of the dry rubber. The elastic Gibbs energy is then:

$$\Delta G_{el} = \left(\frac{N_{el} k_b T}{2} \right) \left(3\alpha^2 - 3 - \ln \alpha^3 \right) \tag{7.16}$$

where N_{el} is the number of elastically effective chains between crosslinks in the rubber network, and the logarithmic term accounts for the entropy change due to the volume change. The excess chemical potential of the solvent is then:

$$\mu_1 - \mu_1^0 = \left(\frac{\partial \Delta G_{mix}}{\partial N_1} \right)_{T,N_2} + \left(\frac{\partial \Delta G_{el}}{\partial \alpha} \right) \left(\frac{\partial \alpha}{\partial N_1} \right)_{T,N_2} . \tag{7.17}$$

The swelling ratio is equal to the reciprocal of the volume fraction of the network, since the initial state is a dry rubber: $\alpha^3 = 1/\phi_2$. Because the network is macroscopic, the ratio of the partial volumes of the network and the solvent is effectively infinity. The excess chemical potential is calculated to be:

$$\mu_1 - \mu_1^0 = k_b T \left(\ln \left(1 - \phi_2 \right) + \phi_2 + \chi \phi_2^2 + \left(\frac{N_{el} v_1}{V_0} \right) \left(\phi_2^{1/3} - \left(\frac{\phi_2}{2} \right) \right) \right). \tag{7.18}$$

If the gel is exposed to excess solvent, the network will continue to swell until the excess chemical potential vanishes. This is called "swelling equilibrium."

The volume fraction of the network at swelling equilibrium depends on the number density of elastically effective chains in the dry network and the solvent quality as measured by the Flory χ parameter. In general, the volume fraction at swelling equilibrium must be calculated numerically, since Equation 7.18 cannot be solved analytically for $\mu_1 - \mu_1^0 = 0$. However, if the volume fraction at swelling equilibrium is low, the equation can be solved approximately. The result is called the Flory–Rehner equation:

$$\phi_2^{-5/3} = \left(\frac{V_0}{N_{el}v_1}\right)\left(\frac{1}{2} - \chi\right). \tag{7.19}$$

The importance of this equation lies in the fact that it is difficult to actually measure the number density of elastically effective chains in the dry network. The measurement of swelling equilibrium in a very good ($\chi = 0$) solvent is a standard technique for characterizing rubber samples.

7.4 Light scattering from concentrated solutions

Although it is very difficult to prepare optically clean concentrated polymer solutions, there are a few reports of such measurements. Since the characteristic lengths in the solution are now small, no structural information is obtained directly from light scattering. The absolute excess scattering factor of the solution, $S(0,c)$, can be calculated from the Einstein theory of light scattering by continuous liquid mixtures. The key experimental quantities are the derivative of the refractive index with respect to concentration, dn/dc, and the derivative of the osmotic pressure with respect to concentration, $d\pi/dc$. The result is given in Chapter 5 as:

$$S(0,c) = \frac{RT}{M(d\pi/dc)}. \tag{7.20}$$

For the Flory–Huggins theory, the solution structure factor is given by

$$S(0,\phi_2) = \left[\frac{x}{1-\phi_2} + 1 - x - 2x\chi\phi_2\right]^{-1}. \tag{7.21}$$

7.5 Real solutions and the Flory–Orwoll theory

Even though the Flory–Huggins theory predicts an upper critical solution temperature and allows a qualitatively correct phase diagram to be calculated, it cannot predict the experimentally observed lower critical solution temperature observed for virtually all polymer solutions. The fundamentally incorrect assumption in the theory is that the volume of mixing of the solution is assumed to be zero. To remedy this problem and improve the predictive power of the theory of concentrated solutions, a full theory for the Gibbs energy as a function of temperature, pressure, and composition must be formulated. An attempt to achieve this goal was made by Orwoll and Flory.[33]

The first task is the creation of an equation of state for a pure liquid that depends on both pressure and temperature. Flory[18] chose a form of corresponding-states theory that can be written in terms of the reduced variables: $\tilde{T} = T / T^{*}$, $\tilde{P} = P / P^{*}$, $\tilde{v} = v / v^{*}$, where the superscripted variables are to be determined experimentally. The equation of state for such liquids is given by:

$$\tilde{P}\tilde{v} / \tilde{T} = \tilde{v}^{1/3} / \left(\tilde{v}^{1/3} - 1\right) - 1 / \tilde{v}\tilde{T} . \tag{7.22}$$

Measurements of the function $v(P,T)$ allow the calculation of the characteristic starred parameters. Equation-of-state data must be obtained for both liquids in the mixture.

The solution thermodynamics is formulated in terms of reduced thermodynamic functions for the mixture. A new parameter similar to the Flory χ is introduced: X_{12}. To account for molecules of different size and shape, a site fraction is introduced for intermolecular interactions: θ. The polymer is divided into r isometric parts, with size chosen so that one isometric unit corresponds to one solvent molecule. For an oligomeric solvent, more than one isometric unit can be assigned to the solvent. Each isometric unit is characterized by s intermolecular contact sites. Corresponding to each intermolecular contact site there is an energy of interaction: $-\eta$. For a two-component liquid mixture, an exchange interaction parameter is defined as:

$$\Delta\eta = \eta_{11} + \eta_{22} - 2\eta_{12} . \tag{7.23}$$

The characteristic pressure and temperature for the solution are then given by:

$$P^{*} = \phi_{1}P_{1}^{*} + \phi_{2}P_{2}^{*} - \phi_{1}\theta_{2}X_{12}$$

$$T^{*} = \frac{P^{*}}{\phi_{1}P_{1}^{*} / T_{1}^{*} + \phi_{2}P_{2}^{*} / T_{2}^{*}} \tag{7.24}$$

where:

$$X_{12} = s\Delta\eta / 2v^{*2} . \tag{7.25}$$

The equation of state for the liquid mixture is taken to be the same as Equation 7.22, with the new characteristic parameters for the mixture. The size parameter, r, is also calculated for the mixture as the total number of isometric units divided by the total number of molecules:

$$\tilde{r} = \left(r_1 N_1 + r_2 N_2 \right) / N . \tag{7.26}$$

The heat of mixing can now be expressed as:

$$\Delta H_{mix} = \tilde{r} N v^* \left[\phi_1 P_1^* \left(1 / \tilde{v}_1 - 1 / \tilde{v} \right) + \phi_2 P_2^* \left(1 / \tilde{v}_2 - 1 / \tilde{v} \right) + \left(\phi_1 \theta_2 / \tilde{v} \right) X_{12} \right] . \tag{7.27}$$

If the reduced volume of the solution differs from the reduced volumes of the two pure liquids, an equation-of-state contribution to the heat of mixing will be observed. The dramatic effect of volume change upon mixing on the measured heat of mixing can be displayed for n-hexane/n-hexadecane mixtures. The data and calculations are shown in Figure 7.2. At lower temperatures, the heat of mixing is positive and goes through a maximum at intermediate compositions. At high temperatures, the heat of mixing is negative and goes through a minimum at intermediate compositions. At intermediate temperatures the heat of mixing changes sign as a function of composition!

The volume change of mixing is indeed negative for this system. The value increases in magnitude with temperature. The theory produces a good qualitative representation of the data, but it overestimates the magnitude. Alkane mixtures change local structure with composition due to the correlations in orientation of the n-hexadecane molecules. The Flory–Orwoll theory assumes random mixing with no change in liquid structure.

Chemical potentials for the solvent relative to the pure liquid can be obtained from vapor pressure or vapor sorption measurements. The characteristic molar volumes for each component are needed for the expression of the excess chemical potential, $\overline{V}_1^* = N_A v_1^*$:

$$\mu_1 - \mu_1^0 = RT \ln \phi_1 + RT \phi_2 \left(1 - (\overline{V}_1^* / \overline{V}_2^*) \right) + \theta_2^2 \overline{V}_1^* X_{12} / \tilde{v}$$
$$+ \overline{V}_1^* P_1^* \left\{ 3 \tilde{T}_1 \ln \left[\left(\tilde{v}_1^{1/3} - 1 \right) / \left(\tilde{v}^{1/3} - 1 \right) \right] + 1 / \tilde{v}_1 - 1 / \tilde{v} \right\} . \tag{7.28}$$

Equation-of-state terms contribute to the chemical potential as well. Application of this theory to alkane mixtures overestimates the chemical potential. The reason is associated with the entropy of mixing due to the changing orientational order of the solution.

The theory does predict lower critical-solution temperatures. The numerical value is not quantitative for mixtures of alkanes with polyethylene, but the inclusion of equation-of-state terms does qualitatively reproduce the coexistence curve for such systems.

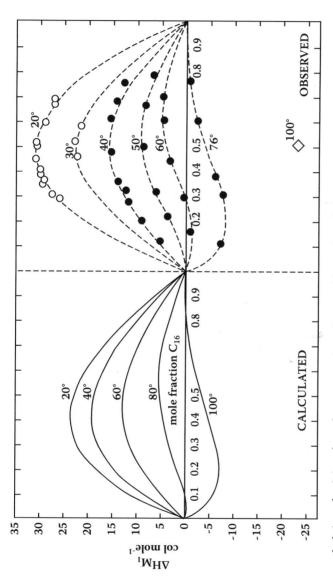

Figure 7.2 Molar enthalpies of mixing for *n*-hexane + *n*-hexadecane. (From Orwoll, R.A. and Flory, P.J., Thermodynamic properties of binary mixtures of n-alkanes, *J. Am. Chem. Soc.*, 89, 6814, 1967. With permission.)

7.6   Diffusion in concentrated solutions

Self-diffusion becomes much slower in high-molecular-weight concentrated solutions. A theory for this process has been developed by deGennes.[9] A chain of N monomers in a semidilute or concentrated solution can be represented as a concatenation of blobs of rms length ξ. The chain is constrained in a tube of other chains of length $L = (N/g)\xi$, where g is the number of monomers per blob. The chain itself has a mean-squared end-to-end length $\langle R^2 \rangle = (N/g)\xi^2$.

The chain diffuses along its tube in a Brownian motion determined by the friction of the chain in the tube. The local friction per blob is given by:

$$f_{blob} = 6\pi\eta_{local}\xi \tag{7.29}$$

and the total friction by $(N/g)f_{blob}$. A tube diffusion coefficient is defined as:

$$D_{tube} = k_b T / (N/g) f_{blob} \sim L^2 / \tau_{rep} . \tag{7.30}$$

The time necessary to diffuse a mean-squared distance along the tube of L^2 is called the reptation time:

$$\tau_{rep} \sim \frac{N^3 \xi^3 6\pi\eta_{local}}{g^3 k_b T} . \tag{7.31}$$

The self-diffusion coefficient can then be expressed as a scaling relation:

$$D_s(c, N) \sim \frac{\langle R^2 \rangle}{\tau_{rep}} \sim \frac{g^2 k_b T}{N^2 \xi 6\pi\eta_{local}} . \tag{7.32}$$

The number of monomers per blob is proportional to $c\xi^3$. And for a good solvent, the screening length is proportional to $c^{-(3/4)}$. This means that the self-diffusion coefficient scales as:

$$D_s(c, N) \sim T / N^2 c^{7/4}\eta_{local} . \tag{7.33}$$

The local viscosity also depends on concentration and temperature, but the details depend on the polymer and the solvent. Experiments carried out with high-molecular-weight chains as a function of chain length verify the $1/N^2$ dependence for the self-diffusion coefficient in well-entangled solutions. The concentration dependence is much harder to verify because of the issue of local viscosity, but the decrease of the self-diffusion coefficient with concentration is dramatic for all solutions in this regime.

The mutual-diffusion coefficient can also be calculated using Equation 6.21. The numerator is obtained from Equation 7.21, and the denominator is given by:

$$\Xi = (cN_A / M) 6\pi\eta_{local} \xi (N / g) .$$
(7.36)

The concentration dependence of $D_m(c)$ in the concentrated regime depends on the solvent quality and the local viscosity. It is often observed to go through a maximum with concentration due to a large increase of local viscosity with concentration. However, if the polymer is well above its glass-transition temperature, the mutual-diffusion coefficient can even increase throughout the concentrated regime. Solutions of poly(dimethyl siloxane) in dioxane exhibit a continuously increasing mutual-diffusion coefficient.

7.7 Viscoelasticity in concentrated polymer solutions

While all liquid media are viscoelastic, concentrated polymer solutions are characterized by an especially rich distribution of relaxation processes. In order to consider this phenomenon, the pure shear deformation will be employed. Consider an equilibrium concentrated polymer solution and identify the absolute location of every atom in the sample. Now deform the sample such that the new x-coordinate of each atom is increased by a constant multiple of its y-coordinate. This affine deformation might be difficult in the laboratory, but it can be executed for a computer sample of a concentrated polymer solution. Now allow the sample to evolve in time in contact with a heat bath. After a long time, the sample will reach equilibrium, with the new shape determined by the initial shear deformation.

The sample as a whole is characterized by a shear-relaxation modulus $G(t)$ that is equal to the instantaneous stress induced in the sample divided by the fixed shear deformation. It is convenient to represent the actual shear-relaxation modulus in terms of a distribution of relaxation times $\rho(\tau)$. The relationship can then be expressed as:

$$G(t) = G(0) \int_0^\infty \rho(\tau) \exp(-t / \tau) d\tau .$$
(7.37)

There will be relaxation processes on many time scales. The most rapid changes will occur between chemically bonded atoms. Chemical bonds that have been stretched or compressed will relax to some excited vibrational level and eventually will thermalize and return to the ground vibrational state. Bond angles will also thermalize and return to the ground vibrational state. Internal torsional angles will initially thermalize within the rotational isomeric state that has been achieved after the deformation. The internal relaxations will release heat into the sample and will lead to a temperature that is out of equilibrium with the heat bath. Heat will then flow back into the heat bath until temperature equilibrium is reached again. The processes described in this paragraph span time scales from picoseconds for the vibrational state relaxations to tens of picoseconds for the initial temperature relaxation.

The chain molecule solution will be characterized by a set of rotational isomeric states. After the initial thermalization, this set will still be in a nonequilibrium state due to the overall shear deformation. For the internal torsional angle distribution to relax to equilibrium, it is necessary for the solution liquid structure to relax and the solvent molecule distribution to reach equilibrium. In a concentrated polymer solution, these processes are highly coupled. Rotational isomeric state changes depend on both internal potentials and the local viscosity and are often in the nanosecond range, well above the glass transition for the solution. Total stress relaxation cannot occur any faster than the chains can change their local rotational isomeric states.

Once the solvent molecules have reached temperature and stress equilibrium, the stress on the sample is borne entirely by the entangled chains. The initial deformation process produces an ensemble of chains that are individually out of equilibrium and a network of entanglements that is out of equilibrium. At sufficiently long time, the concentrated solution reaches a state of vanishing stress, with an equilibrium distribution of individual chain conformations and an equilibrium ensemble of chain entanglements. The deformed chains between entanglements will relax to a sequence of equilibrium statistical segments. This process occurs on a time scale determined by the local viscosity and the size of the local correlation length for the solution. The time scales are between nanoseconds and microseconds for typical concentrated solutions well above the glass transition. For the overall chain to release the entanglements, it must undergo self-diffusion on a time scale denoted above as the reptation time. This time depends on the local viscosity, the size of the statistical subunits, and the chord length of the chains, L. For sufficiently long chains, the reptation time for the solution can exceed 1000 sec, even for solutions well above the glass transition, and the concentrated solution behaves as a gel on short time scales.

This overview of the range of time scales that are important for the description of a concentrated polymer solution emphasizes the complexity of these systems. More than 15 decades of time are often involved in the dynamics of concentrated polymer solutions. The dynamics can be organized into groups of relaxation times associated with specific types of relaxations. The local fluid structure and local rotational isomeric state dynamics are most closely associated with the phenomenon known as the glass transition and will be considered in more detail in Chapter 8.

The detailed ideas of deGennes[9] outlined above for the self-diffusion of linear chains in a concentrated solution lead to an explicit expression for the reptation time (Equation 7.31). These same ideas can be applied to the relaxation of the local chain segments of size ξ called blobs. They will relax on a time scale that scales as:

$$\tau_{\text{blob}} \sim \xi^3 \eta_{\text{local}} / T .$$

(7.38)

The ratio of the local blob relaxation time to the reptation time depends on the third power of the chain length and can become very large. In the long-chain limit, the "pseudogel" character of the solution is very evident. One question that must arise is the lower limit of the intramolecular length that determines the cooperative relaxation of the backbone bonds. The persistence length, a, of the chain is determined primarily by intramolecular structure and rotational potentials and is only weakly dependent on the solvent in an organic solvent. The strong effect of the ionic strength and pH of an electrolyte solution on the persistence length of a linear polyelectrolyte chain will be considered in Chapter 10. For a nonelectrolyte solution, there should be a limiting relaxation time for the chain on length scales comparable to the persistence length that depends only on the local viscosity, the temperature, and a^3.

Another experimental quantity that also reflects the full dynamics of the solution is the creep compliance, $J(t)$. A fixed stress is applied to the sample, and the strain is followed as a function of time. It is customary to represent the creep compliance in the form:

$$J(t) = J_g + J_R(t) + t / \eta$$

(7.39)

where the initial term reflects the glassy compliance associated with the local fluid relaxations and torsional isomeric state dynamics. The recoverable compliance increases with time to a limiting steady-state value, J_e^0. The long-time behavior is determined entirely by the shear viscosity for the solution. A good approximation for the shear viscosity of a well-entangled concentrated solution is:

$$\eta = \tau_{\text{rep}} / J_e^0 .$$

(7.40)

The value of the steady-state compliance due to the entanglement network depends on the polymer and the concentration. For bulk liquid polymer, there exists a limiting molecular weight below which the liquid does not display a rubberlike compliance, M_e. The corresponding compliance for the solution can be expressed as:

$$J_e^0 \approx J_N^0 \sim M_e \rho / c^2 RT$$

(7.41)

where the density of the bulk polymer must also be known. The solution viscosity then scales as:

$$\eta \sim \frac{N^3 c^{7/2} \eta_{\text{loc}}}{M_e} .$$

(7.42)

for a good solvent. Concentrated solution viscosities increase very dramatically with chain length and concentration.

The concept of a local viscosity has been very useful in the interpretation of the properties of polymer solutions. One series of experiments, which probed the concentration dependence of the local viscosity on a length scale comparable to the solvent, observed the rotational relaxation time for styrene monomer during thermal polymerization using depolarized Rayleigh scattering. The rotational relaxation time increased only slowly well into the concentrated regime. The overlap concentration was near 1% polymer, and the styrene relaxation time continued to increase linearly with time until the concentration exceeded 50% for polymerization at 363 K. As solvent molecules are converted to polymer molecules, the total volume of the solution decreases at constant temperature, and the local viscosity increases accordingly. The relationship between viscosity and volume is explored in more detail in Chapter 8. The rotational relaxation time for pure styrene was determined as a function of temperature and viscosity. The calculated relaxation time depends linearly on the quantity η/T. The linear dependence of the relaxation time on polymerization time at constant temperature (shown in Figure 7.3) indicates that the local viscosity increases linearly with time as the styrene is converted thermally to polystyrene.

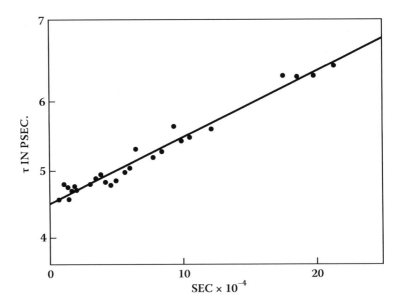

Figure 7.3 Rotational relaxation time of styrene monomer during thermal polymerization at 363 K. The relaxation time is linearly dependent on the quantity η/T, so that the local viscosity increases linearly at constant temperature as the styrene converts to polystyrene. (From Alms, G.R. et al., A study of the thermal polymerization of styrene by depolarized Rayleigh scattering spectroscopy, *J. Chem. Phys.*, 70, 2145, 1979. With permission.)

chapter eight

Structure and properties of polymers in the pure amorphous liquid state

8.1 Introduction

Polymers can also exist as pure amorphous liquids. Assertions to the contrary have now been largely abandoned. Since the screening lengths are comparable to the size of the monomer units, polymer chains obey Gaussian statistics on length scales comparable to the size of the statistical subunits and the whole molecule. Neutron-scattering experiments have verified this conclusion. The insight that this would be the case was presented by Flory[15] and deGennes.[9]

While the Flory-Orwoll theory for the equation of state of bulk polymers can be employed, a simpler free-volume equation of state[10] is often used in practice. The details of this theory and its application to the description of the viscosity of low-molecular-weight polymer liquids is presented in Section 8.2.

The shear viscosity of low-molecular-weight polymeric liquids is often observed to scale linearly with the molecular weight as long as the chain length is long enough to yield the limiting local structure for the liquid and short enough to avoid entanglements. This experimental result can be rationalized in terms of the Rouse theory[12] of polymer chain dynamics and is presented in Section 8.3.

As polymer liquids are cooled or compressed, the local viscosity increases. If the average local relaxation time of the system exceeds the time allowed for equilibration, the liquid follows a nonequilibrium path. This fact leads to the phenomenon of the glass transition. The phenomenology of the glass transition is presented in Section 8.4.

The liquid state near the glass transition can be described in terms of a distribution of relaxation times. The temperature and pressure dependence of these relaxation times is discussed in Section 8.5.

8.2   Free volume and viscosity

Doolittle[10] proposed that the specific volume of a liquid as a function of chain length could be empirically represented as:

$$\ln v = \ln(1/\rho) = a / M + b \tag{8.1}$$

where a and b are empirical functions of temperature:

$$\ln a = C_1 T + C_2 ;$$
$$b = C_3 T^{C_4} + C_5 . \tag{8.2}$$

If the specific volume is extrapolated to absolute zero, the result for n-alkanes was found to be $v_0 = \exp(10 / M)$. The free volume of the liquid is then defined to be $v_f = v - v_0$.

The free volume was found to be an empirically useful quantity in the correlation of measured shear viscosities:

$$\eta = A \exp\left(B / \left(v_f / v_0\right)\right) \tag{8.3}$$

where A and B are empirical constants for each liquid. This relationship was very effective in describing n-alkane viscosities. The concept of free volume appears in many discussions of the viscoelastic properties of bulk polymers.

8.3   Viscosity of low-molecular-weight chain liquids

Once the chains are long enough to reach the asymptotic characteristic ratio discussed in Chapter 2, it is found that the measured shear viscosity increases linearly with molecular weight for many polymer liquids. This phenomenon can be explained in terms of the dynamics of individual polymer chains as long as the system is below the entanglement limit. The basic theory is due to Rouse and is described in detail by Ferry.[12]

The chain is modeled as a system of beads and springs undergoing Brownian motion in a viscous medium. The other polymer chains provide the viscous medium for any individual chain. The inherent dynamics can be represented in terms of N relaxation modes, where N is the number of statistical subunits in the chain. The shear relaxation modulus $G(t)$ is given by:

$$G(t) = \left(\frac{\rho R T}{M}\right) \sum_{p=1}^{N} \exp(-t / \tau_p) ; \tag{8.4}$$

where the relaxation times are given by:

$$\tau_p = \frac{6\eta M}{\pi^2 p^2 \rho RT} .$$
(8.5)

There is a certain circularity in this relationship, since the shear viscosity is defined as the integral of the shear relaxation modulus. A useful relationship for the shear viscosity is in terms of the relaxation modulus and the average shear relaxation time:

$$\eta = \left(\frac{\rho RT}{M} \right) N \langle \tau \rangle = \left(\frac{\rho RT}{M} \right) \int_0^\infty \sum_{p=1}^N \exp\left(-t / \tau_p \right) dt .$$
(8.6)

This means that each molecule contributes N terms of $k_b T$ to the relaxation modulus. Each relaxation mode depends on the overall molecular weight of the chain to the first power, explicitly, and on the viscosity itself, which depends on the first power of the molecular weight.

8.4   Phenomenology of the glass transition

As polymer liquids are cooled or compressed, the viscosity increases. Eventually the viscosity is high enough that equilibrium is not reached upon further changes in a reasonable length of time. The liquid then ceases to behave as an equilibrium fluid and is called a glass. However, the glass is not a static system and continues to evolve in time. Isothermal annealing continues to occur until the liquid reaches its equilibrium volume. The hysteresis associated with changes near the glass transition is the defining property of the phenomenon. The path followed by the liquid during changes in temperature or pressure determines the microscopic state of the fluid. Non-equilibrium liquids can have the same actual volume but be in different microscopic states. This picture of the liquid state is elaborated in the discussion below.

Careful measurements of the specific volume of polymer liquids as a function of thermal history have been carried out by Kovacs.[24] One class of experiments subjected the liquid to a series of cooling trajectories. The cooling rate is varied over many decades in time: $q = dT/dt$. The specific volume decreases along the equilibrium liquid line until it can no longer adjust to the temperature changes. Over an interval in temperature, the apparent thermal-expansion coefficient decreases until it reaches a value more consistent with a solidlike behavior. The cooling trajectory can be characterized by a fictive temperature, T_f, that is obtained by extrapolating the liquid and glass lines until an intersection temperature is reached. The value of the fictive temperature is an explicit function of the cooling rate for each liquid.

The most rapid cooling rate is limited by the thermal conductivity of the organic liquid and is on the order of 1 K/s. The slowest rate is limited by the stability of the apparatus and the patience of the experimenter and is often in the range 10^{-5} K/s. The fictive temperature changes by from 10 to 20 K over this range of cooling rates for organic materials near 300 K.

Another thermal history consists of thermal equilibration at a temperature above the glass-transition range followed by rapid quench to a temperature within the glass-transition range. Isothermal annealing is then followed until equilibrium is reached. A relaxation function is defined as:

$$\phi_V\left(T,t\right)=\left(v\left(T,t\right)-v\left(T,\infty\right)\right)/v\left(T,\infty\right). \tag{8.7}$$

As the quench temperature is lowered, the initial value of the specific volume exceeds the equilibrium volume by an increasing magnitude. The average relaxation time increases rapidly as the quench temperature is lowered. A typical volume relaxation curve is shown in Figure 8.1.

The shape of the curves is highly nonexponential. One obvious reason is that the rate of relaxation depends on the free volume that is changing during the volume relaxation. However, attempts to explain the observed relaxation functions in terms of a single volume-dependent relaxation time have not been successful. Because equilibrium liquids can be obtained at

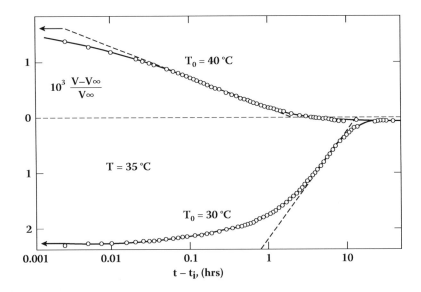

Figure 8.1 Volume relaxation curves for a downward and an upward quench near the glass-transition temperature. (The time axis is logarithmic, and the polymer is poly[vinyl acetate]). (From Kovacs, A.J., Glass transition in amorphous polymers: a phenomenological study, *Adv. Polym. Sci.*, 3, 394, 1963. With permission.)

any temperature within the glass-transition range, it is possible to start in the equilibrium state and rapidly increase the temperature to an annealing temperature still within the hysteresis range and then follow the volume increase until the new equilibrium volume is achieved. A typical relaxation function of this type is shown in Figure 8.1. Now the volume increases monotonically, and the rate of relaxation might be expected to follow. A convenient expression for the relaxation rate is:

$$\text{rate} = 1 / \tau = -(1 / (v - v_\infty))(dv / dt) . \tag{8.8}$$

In fact, the rate of relaxation is highest at the start of the volume expansion, decreases to a minimum, and then increases until the end of the experiment. This behavior can be explained in terms of a distribution of relaxation times for volume fluctuations.

The absolute need for a distribution of relaxation times can be illustrated by a more complicated thermal history. The sample is annealed to equilibrium above the glass-transition interval and then quenched rapidly to a temperature deep within the glass-transition region. It is isothermally annealed until its volume achieves a value consistent with a glass with a fictive temperature in the middle of the glass-transition interval. The partially annealed glass is then rapidly heated to this fictive temperature, where the sample achieves a volume exactly equal to its equilibrium value. The volume then increases spontaneously to a maximum value and finally relaxes back to its equilibrium volume. This trajectory is called the "memory effect" and is shown in Figure 8.2.

The picture of the liquid that is consistent with these observations includes a distribution of regions within the equilibrium (or nonequilibrium) liquid that differ in their local density and mobility.[11] Any compressible fluid at finite temperature will have density fluctuations, and the local mobility depends on the local free volume. The distribution of local densities will be a function of temperature and thermal history. When a sample in the glassy state is rapidly heated, the initial thermal energy is channeled into local intermolecular vibrational motions. This increases the local specific volume throughout the sample. In regions of large specific volume, the vibrational energy can be rapidly converted into a lower density by nucleating free volume. The local volume increase leads to a local temperature decrease, since it must occur against the attractive intermolecular potential. The system rapidly evolves into a system with local pressure and temperature gradients. The excess free volume created in the least dense regions must diffuse into the denser regions to reach equilibrium. Since the annealing is carried out isothermally, the sample can continue to absorb energy from the heat bath as it expands. The realization that both the volume and the enthalpy must be followed to understand the trajectory is a key concept. Near the end of the trajectory, the enthalpy is approaching its equilibrium value, and the least dense regions are in equilibrium with one another.

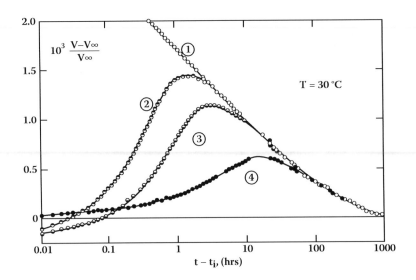

Figure 8.2 Volume as a function of time after memory effect thermal history. Curve 1 is a pure quench. Curves 2 to 4 correspond to quenches and annealing at 10, 15, and 25°C, respectively. (From Kovacs, A.J., Glass transition in amorphous polymers: a phenomenological study, *Adv. Polym. Sci.*, 3, 394, 1963. With permission.)

One of the most common experimental techniques used to study the glass transition is differential scanning calorimetry (DSC). The sample is heated at a constant rate, and the energy flow necessary to maintain this constant heating rate is monitored as a function of time. The elapsed time is then converted to a temperature, and the practical heat capacity ($C_p = dH/dt$) is plotted against calculated temperature. An "ideal" heating curve for a glassy sample exhibits a lower output along the glassy part of the curve and a smooth rise over a narrow temperature interval to a limiting liquid value for the practical heat capacity. Actual curves can be more complicated.

Because the heating curve depends on the actual state of the sample before heating, the thermal history of the sample is important. A typical thermal history includes a rapid cooling path followed by a slower heating path. As the sample is heated, it continues to anneal toward its equilibrium volume until the equilibrium line is reached. Because the sample now has a lower fictive temperature than it did during cooling, the heating cycle will produce a liquid with a specific volume lower than the equilibrium value for a small interval past the equilibrium line. The volume then rapidly rises to reach its equilibrium value, as the mobility is rapidly increasing above the glass transition interval. As this occurs, the energy necessary to produce the constant heating rate exceeds that needed to maintain the heating rate for a pure equilibrium liquid. The practical heat capacity will display a weak maximum before decreasing to its equilibrium liquid magnitude.

An even more dramatic maximum can be produced by isothermally annealing the sample before reheating. Much lower fictive temperatures can

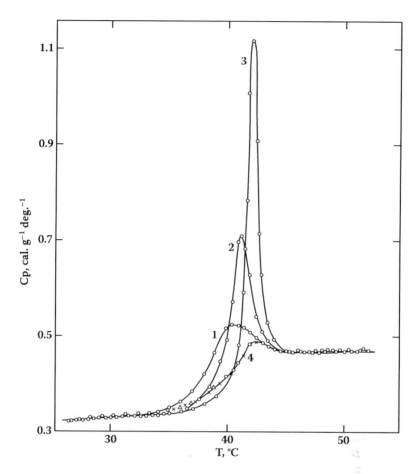

Figure 8.3 DSC curves for samples with different thermal histories. Well-annealed samples display a substantial overshoot peak. Curve 4 was for a quenched and immediately reheated sample. Curves 1 to 3 were for increasing annealing up to 7 days for curve 3. (From Kovacs, A.J., Glass transition in amorphous polymers: a phenomenological study, *Adv. Polym. Sci.*, 3, 394, 1963. With permission.)

then be obtained. During heating, the sample will continue along the glass line far beyond the equilibrium liquid line, and the eventual rise to the equilibrium line is much steeper. The practical heat capacity can exhibit an overshoot peak many times the intrinsic jump from the glassy to the liquid heat capacity. An example of these DSC curves is shown in Figure 8.3.

8.5 Temperature and pressure dependence of relaxation near the glass transition

The shear-relaxation modulus can be expressed in terms of a distribution of relaxation times $\rho(\tau)$, as in Equation 7.37. While there is no fundamental reason why the shape of the distribution should be independent of temperature and pressure, it is sometimes observed that this is the case. When it is true, the material is referred to as rheologically simple. The ratio of the relaxation times at two temperatures is called the shift factor, a_T, and is often expressed in the form[12]:

$$\log a_T = -\frac{c_1^0 \left(T - T_0\right)}{c_2^0 + T - T_0} \tag{8.9}$$

where T_0 is an arbitrary reference temperature.

It has been proposed that the shift factor is a function of fractional free volume, $f = v_f / v$:

$$\log a_T = \frac{B}{T} \left(\frac{1}{f} - \frac{1}{f_0}\right). \tag{8.10}$$

This is similar to the Doolittle expression for viscosity. It is further assumed that near the glass transition:

$$f = f_0 + \alpha_f \left(T - T_0\right). \tag{8.11}$$

If Equation 8.11 is used in Equation 8.10, the form of Equation 8.9 is obtained with:

$$c_1^0 = B / 2.303 f_0 \qquad c_2^0 = f_0 / \alpha_f . \tag{8.12}$$

Another functional form that has been found useful in the description of relaxation times and viscosities near the glass transition is:

$$\tau = \tau_0 \exp\left(\frac{A + BP}{T - T_\infty}\right). \tag{8.13}$$

Relaxation times as a function of pressure are often found to obey Equation 8.13. The pressure dependence can be rationalized using Equation 8.10 by adding a term that depends on the compressibility of the free volume.

Since the relaxation times are often observed to be continuous functions of temperature and pressure, it is not clear how to define the glass transition for an equilibrium fluid. It is clear that there is no unique temperature that defines the phenomenon, and most practical work adopts some arbitrary criterion. The temperature, T_∞, sets a lower bound, since the relaxation times extrapolate to large values at this temperature. Temperatures within the range observed in standard DSC curves are often chosen, but even here the protocol for extracting a T_g from the measured practical heat capacity curves is not clear. Some workers prefer the fictive temperature, but again, the value of T_f is an explicit function of thermal history. Handbooks of polymers have tables of glass-transition temperatures. It is well to look deeper and find out what criterion has been chosen.

The measured shape of the shear-relaxation modulus depends on the actual material. However, the part of the relaxation from the glassy initial modulus down to the modulus associated with the Rouse modes (described above) is often much more universal. The shape is often well represented by the form:

$$G(t) = G(0) \exp\left(-\left(\frac{t}{\tau}\right)^\beta\right)$$ (8.14)

where the fractional power exponential is typically in the range 0.2 to 0.5. This functional form is often taken as the "signature" of the glass-transition phenomenon. Equilibrium liquids near the glass transition have volume relaxation times in the range 0.01 to 1000s and highly nonexponential relaxation functions. The relaxation times shift according to Equation 8.13. The shape of the distribution of relaxation times corresponding to the fractional power exponential functions has been calculated numerically.[26] Experimental observation of the distribution of relaxation times has also been shown to produce good agreement with this function. Depolarized Rayleigh scattering from bulk polymer liquids is dominated by rotational motion of very local chain elements. The experimental distribution of relaxation times for polystyrene near its glass-transition region is shown in Figure 8.4.

Another representation of the dynamics of the glass transition is the creep compliance. The functional form of the initial part of the creep compliance near the glass transition is often found to be:[12]

$$J(t)_{\text{initial}} = J_g + Bt^{1/3} + t/\eta .$$ (8.15)

Many materials have been studied both as equilibrium liquids and as nonequilibrium systems near the glass transition.[39] Even dry cheese obeys Equation 8.15!

The discussion in Section 7.7 also made mention of processes at very short times. Even when the volume relaxation time is in the millisecond time range, there are often molecular processes taking place on picosecond, nanosecond, and microsecond time scales. Three convenient experimental techniques to study these processes are dynamic light scattering, mechanical relaxation, and dielectric relaxation.

Dynamic light scattering in bulk polymers is reviewed by Patterson.[34] The calculated distribution of relaxation times obtained from dynamic light scatter-

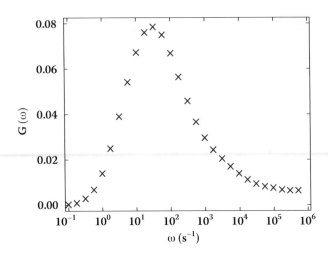

Figure 8.4 Experimental distribution of relaxation frequencies ($\omega = 1/\tau$) for polystyrene near its glass transition. (From Carroll, P.J. and Patterson, G.D., The distribution of relaxation frequencies from photon correlation spectroscopy near the glass transition, *J. Chem. Phys.*, 82, 9, 1985. With permission.)

ing reveals up to three groups of relaxation processes.[37] The longest relaxation time group corresponds to the primary glass-transition processes that dominate the volume relaxation and is often called the α-relaxation. The next-fastest group is associated with (a) local intramolecular relaxation processes that involve coupling to the fluid structure and (b) intermolecular relaxation processes that lead to only partial fluid structure relaxation. They are often called secondary or β-relaxations. Even when the primary relaxations are in the glass-transition regime, secondary relaxations often have relaxation times in the nanosecond to microsecond range. Since many polymers have flexible side chains, the dynamics of these chemical structures can only be weakly coupled to the fluid structure. A distribution of relaxation times obtained for poly(n-hexyl methacrylate) is shown in Figure 8.5. The primary and second relaxations are highly overlapped in this material, and there is evidence of the long-time part of a faster relaxation process. However, there are many materials where a clear separation between the α- and β-relaxations is evident.

Mechanical relaxation consists of measuring the frequency dependence of the complex mechanical modulus. For the shear modulus, the relationship of interest is:

$$G^*\left(\omega\right) = \omega \int_0^\infty G\left(t\right)\exp\left(i\omega t\right)dt = G' + iG'' . \tag{8.16}$$

For an equilibrium liquid, the real part of the complex shear modulus grows from zero at low frequencies to $G(0)$ for high frequencies. The imaginary part goes through a maximum at a frequency determined by the distribution

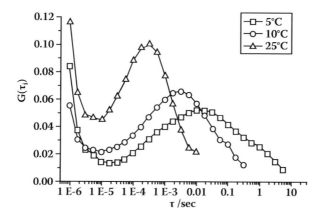

Figure 8.5 Distribution of light-scattering relaxation times for poly(hexyl methacry-late) near its glass-transition range. At the lowest temperature, the primary and secondary relaxations are separating. The long-time part of the gamma relaxation is visible in the short time window of the technique. (From Savin, D.A., Patterson, G.D., and Stevens, J.R., Evidence for the γ-relaxation in the light scattering spectra of poly(n-hexyl methacrylate) near the glass transition, *J. Polym. Sci. B: Polym. Phys.*, 43, 1504, 2005. With permission.)

of shear relaxation times. A typical experimental program obtains the frequency of maximum mechanical loss as a function of temperature and pressure. For the primary glass-transition processes, the frequency of maximum loss is often well represented by:

$$\omega_{max,\alpha} = A \exp\left(-\frac{B}{T - T_\infty}\right). \qquad (8.17)$$

Many polymeric liquids display a maximum in G'' at higher frequencies than those associated with the primary glass relaxation. The secondary maximum can have a relaxation strength (as measured by the value of the distribution of relaxation times) that exceeds the primary glass relaxation strength. The frequencies of maximum loss often obey the relation:

$$\omega_{max,\beta} = A \exp(-B / T). \qquad (8.18)$$

This is consistent with the local intramolecular character of many of these relaxations. It is also observed that the breadth of the distribution of relaxation times associated with the secondary relaxations is often much larger than that associated with the primary glass relaxation.

If some of the chemical bonds are polar, the molecular dynamics can be followed by measuring the complex dielectric constant $\varepsilon^*(\omega) = \varepsilon' + i\varepsilon''$. If there is a polar character to the main-chain bonds, the entire polymer molecule may have a net dipole moment. For these polymers, the dielectric relaxation will have components that reflect the Rouse modes discussed in Section 8.3. For polymers with molecular weight below the limit for chain entanglement, the longest Rouse mode can be determined from the frequency of maximum dielectric loss at low frequencies for liquids far above the glass transition. As long as there are components of the local dipole moment transverse to the chain axis, dielectric relaxation can be used to probe the primary glass relaxation. If there are components of the local dipole moment associated with side chains, dielectric relaxation can be used to probe secondary relaxations. Many studies of this type are summarized by McCrum et al.[27]

At the longest times associated with the shear-relaxation modulus, the disentanglement processes determine the value of the shear viscosity. It is observed experimentally that the shear viscosity depends on chain length as:

$$\eta \sim M^{3.4} . \tag{8.19}$$

Arguments presented in Section 7.7 lead to a cubic dependence on molecular weight. More subtle effects involving local fluctuations and self-knots in chains must be invoked to augment the molecular-weight dependence.

chapter nine

Structure and properties of rodlike polymers in solution

9.1 Introduction

While there is no such thing in nature as a rigid rod polymer, it has been found very useful to consider the properties of such an object in solution. The rod is characterized by its length, L, and its diameter, d. The characterization of such an object in solution is conveniently carried out using light scattering. The theory and some applications of light scattering to the measurement of stiff chains in solution are presented in Section 9.2.

Dilute solutions of rodlike polymers have been treated thermodynamically by Tanford.[40] The theory and a few applications are presented in Section 9.3, but, for long rods, the dilute concentration range is very limited. A limiting concentration can be expressed as $c_L^* = M / N_A L^3$. For rod lengths in excess of 100 nm, the dilute regime almost vanishes.

Consideration of solutions of rodlike polymers at higher concentrations was presented by Flory.[17] The lattice model used so successfully for the theory of concentrated polymer solutions was extended to include semiflexible chains and rods. The theory predicted a miscibility gap between a dilute isotropic solution and an ordered concentrated solution. This theory is presented in Section 9.4. The same solvent quality issues needed to address the thermodynamics of flexible polymer solutions also influence the phase behavior of lyotropic solutions. A description of actual solutions of stiff polymers is also presented.

9.2 Characterization of rodlike polymers in solution

Measurement of the light-scattering structure factor, $S(q)$, for a rodlike particle in solution is a convenient method for characterizing its structure. It is convenient to introduce the dimensionless variable $x = qL$. The functional form is given by Berne and Pecora[3] as:

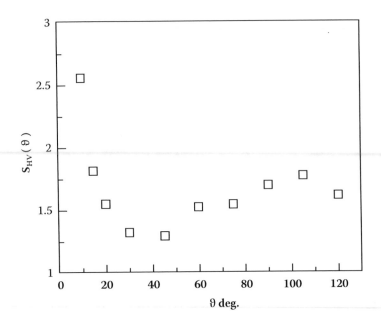

Figure 9.1 Depolarized structure factor for lithium molybdenum selenate in dilute propylene carbonate solution. The measured rod length is 2 μm. (From Patterson, G.D., Ramsay, D.J., and Carroll, P.J., Depolarized light scattering spectroscopy and polymer characterization, *Analytica Chimica Acta*, 189, 57, 1986. With permission.)

$$S(q) = \frac{2}{x}\int_0^x \frac{\sin z}{z}dz - \left(\frac{\sin(x/2)}{x/2}\right)^2. \tag{9.1}$$

While this is an exact mathematical form for an infinitely thin rod, it raises two important issues for real polymers. At higher values of x, the finite diameter of the rod becomes important. Measurement of d is usually carried out using neutron scattering, where higher values of q are available. The structure factor for a thin rod is a monotonically decreasing function of angle. If light scattering is used to obtain the structure factor, the observed angular dependence also contains contributions from the anisotropic polarizability, γ^2, of the rod. This potential "problem" can be turned to advantage, since the depolarized structure factor for a rod has also been reported by Berne and Pecora[3] (Equation 9.2). While there are a large number of terms, a computer makes the calculations straightforward. An actual measured depolarized structure factor for a long rod in solution is shown in Figure 9.1. It is characterized by a clear minimum, from which the rod length can be calculated. The rod length can then be established for any long rod that can be obtained in dilute solution:

$$S_{HV}(\theta) = \gamma^2 \left[(1/2)Y(x) - (3/4)X(x) - (1/4)\cos\theta Z(x) \right]$$

where:

$$X(x) = \left(\frac{\sin x}{x^5}\right) - \left(\frac{\cos x}{x^4}\right) + \left(\frac{\sin x}{2x^3}\right) + \left(\frac{\cos x}{2x^2}\right) - \left(\frac{4}{3x^2}\right) + \left(\frac{1}{2x}\right) \int_0^x \frac{\sin z}{z} dz \quad (9.2)$$

$$Y(x) = \left(\frac{\sin x}{x^3}\right) + \left(\frac{\cos x}{x^2}\right) - \left(\frac{2}{x^2}\right) + \left(\frac{1}{x}\right) \int_0^x \frac{\sin z}{z} dz$$

$$Z(x) = \left(\frac{5\sin x}{x^5}\right) - \left(\frac{5\cos x}{x^4}\right) + \left(\frac{\sin x}{2x^3}\right) + \left(\frac{\cos x}{2x^2}\right) - \left(\frac{8}{3x^2}\right) + \left(\frac{1}{2x}\right) \int_0^x \frac{\sin z}{z} dz$$

The rod length and diameter can also be inferred from dynamic depolarized Rayleigh scattering. The translational and rotational diffusion coefficients can be determined. The depolarized light-scattering correlation function can be expressed as:

$$\phi_{HV}(q,t) = \exp\left(-\left(6\Theta + D_0 q^2\right)t\right). \quad (9.3)$$

The measured diffusion coefficients can then be interpreted in terms of the length and diameter according to the theory of Perrin.[34] It is convenient to define an axial ratio $\rho = d/L$. The translational and rotational diffusion coefficients can be expressed as:

$$D_0 = \frac{k_b T G(\rho)}{3\pi\eta L}$$

$$\Theta = \left(\frac{3k_b T}{2\pi\eta L^3}\right) \frac{\left[(2-\rho^2)G(\rho) - 1\right]}{(1-\rho^4)}$$

where: $\qquad (9.4)$

$$G(\rho) = \frac{\ln\left\{\frac{1 + (1-\rho^2)^{1/2}}{\rho}\right\}}{(1-\rho^2)^{1/2}}$$

Analysis of the dynamic depolarized scattering from the rods shown in Figure 9.1 yielded a rod with a length near 2 μm and an axial ratio less than 0.02.

9.3   Second osmotic virial coefficient

The osmotic second virial coefficient for a dilute solution of solid hard cylinders of length L and diameter d can be evaluated from geometric considerations alone. The presence of another rod whose center of mass is within a distance L restricts the available rotational freedom of the first rod. Calculations by Zimm et al. in a work by Tanford[40] all agree that the second virial coefficient for these objects is given by:

$$B = L\bar{v} / dM .$$

(9.5)

For a fixed molar mass and partial specific volume, the second virial coefficient is inversely proportional to the axial ratio. For a fixed diameter, the second virial coefficient is independent of rod length, since L scales as M.

Actual rodlike polymers in solution are influenced by intermolecular attractions of longer range due to van der Waals and electrolyte effects. An attempt to deal with solvent quality is presented below in connection with the Flory theory of rodlike solutions. Electrolyte solutions are considered in Chapter 10.

9.4   Thermodynamics of rodlike polymer solutions

Flory[17] has presented a theory of rodlike polymers in solution. The solvent is chosen to be a small molecule that defines the lattice geometry. The rod is represented as a linear sequence of x lattice cells where, as usual, x is the ratio of the molar volumes of the polymer and the solvent. The lattice is characterized by a coordination number z. An initial theoretical effort focused on the athermal case. For a mixture of N_1 solvent molecules and N_2 polymer molecules, the canonical ensemble partition function for the isotropic liquid mixture can be expressed as:

$$Q_M = q_1^{N_1} q_2^{N_2} \left[\frac{(N_1 + xN_2)!}{N_1! N_2! (N_1 + xN_2)^{(x-1)N_2}} \right] \left(\frac{z}{2} \right)^{N_2}$$

(9.6)

where the q's are single particle partition functions for the solvent and the rod. The existence of an explicit partition function allows the chemical potential for the solvent and the polymer in the isotropic solution to be derived. Such solutions are known to be isotropic only at very low concentrations.

It is known empirically that solutions of rodlike polymers are highly ordered at higher concentrations. The partition function for a completely orientationally ordered but otherwise random solution is:

$$Q_M = q_1^{N_1} q_2^{N_2} \frac{(N_1 + N_2)!}{N_1! N_2!}. \tag{9.7}$$

Flory devised a scheme to calculate the partition function for a solution with a definite intermediate degree of order along a unique nematic axis. The angle of a particular rod with respect to this axis is denoted ψ_i, and the rod is subdivided into $y_i = x \sin \psi_i$ subrods containing x/y_i segments. The ensemble of N_2 rods can be divided into a distribution of orientations with angle near ψ_k: $\{N_k\}$. The average value of y_i for all the rods can then be calculated and denoted y. The partition function can then be expressed as:

$$Q_M = q_1^{N_1} q_2^{N_2} \left[\frac{(N_1 + yN_2)!}{N_1! N_2! (N_1 + xN_2)^{(y-1)N_2}} \right] \left(\frac{N_2!}{\prod_k N_k!} \right). \tag{9.8}$$

The free energy of mixing for an athermal anisotropic solution can then be derived:

$$\Delta G_M = k_b T \left[N_1 \ln \phi_1 + N_2 \ln \phi_2 - (N_1 + yN_2) \ln \left(1 - \phi_2 (1 - y/x) \right) \right.$$
$$\left. - N_2 \left(\ln \left(xy^2 \right) - y + 1 \right) \right]. \tag{9.9}$$

The value of y for a given value of x and ϕ_2 is obtained by minimizing the free energy of mixing. The condition of equilibrium demands that:

$$\phi_2 = \left(\frac{x}{x-y} \right) \left(1 - \exp(-2/y) \right). \tag{9.10}$$

For large values of x, it can be shown that stable anisotropic solutions exist only for volume fractions higher than:

$$\phi_2^* = (8/x)(1 - (2/x)). \tag{9.11}$$

From these considerations, the chemical potentials of the solvent and the solute in the anisotropic phase can be calculated as a function of x and volume fraction. The chemical potential for the solvent in an athermal anisotropic solution is:

$$\left(\mu_1 - \mu_1^0\right)/RT = \ln\left(1 - \phi_2\right) + \left[\left(y-1\right)/x\right]\phi_2 + 2/y \qquad (9.12)$$

compared with the isotropic solution:

$$\left(\mu_1 - \mu_1^0\right)/RT = \ln\left(1 - \phi_2\right) + \left(1 - 1/x\right)\phi_2 . \qquad (9.13)$$

The corresponding chemical potential for the rods in the anisotropic solution is:

$$\left(\mu_2 - \mu_2^0\right)/RT = \ln\left(\phi_2/x\right) + \left(y-1\right)\phi_2 + 2 - \ln y^2 \qquad (9.14)$$

and in the isotropic solution is:

$$\left(\mu_2 - \mu_2^0\right)/RT = \ln\left(\phi_2/x\right) + \left(x-1\right)\phi_2 - \ln x^2 . \qquad (9.15)$$

A solution phase diagram can then be calculated as a function of x. The result is shown in Figure 9.2. Below a critical value of x, all solutions, including pure liquid solute, are isotropic. For longer chain lengths, there is a gap between the concentration of the isotropic phase and the ordered phase with the same chemical potentials. Miscibility gaps in solutions of rodlike polymers are routinely observed.

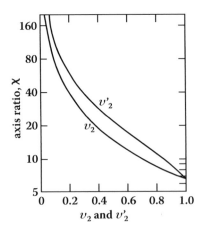

Figure 9.2 Concentration of phases in equilibrium as a function of x for athermal mixtures of solvent and rods. (From Flory, P.J., Phase equilibria in solutions of rod-like particles, *Proc. R. Soc., A*, 234, 73, 1956. With permission.)

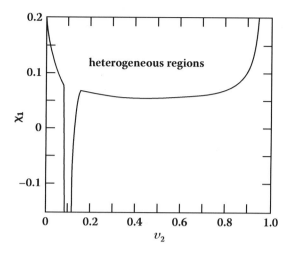

Figure 9.3 Phase diagram for a solution of rods with $x = 100$ as a function of the interactions parameter χ. (From Flory, P.J., Phase equilibria in solutions of rod-like particles, *Proc. R. Soc., A*, 234, 73, 1956. With permission.)

Most solutions of rodlike polymers are not athermal. Heat-of-mixing terms can be added to the chemical potentials, just as in the ordinary Flory–Huggins theory of solutions. When the interaction parameter χ is large enough, the solution will separate into a dilute phase and a concentrated phase, just as for any polymer solution. There will still be a miscibility gap between dilute isotropic solutions and ordered concentrated solutions. A typical phase diagram for a solution of rods with $x = 100$ is shown in Figure 9.3. The concentrated phase is now very concentrated and ordered.

chapter ten

Structure and properties of polyelectrolyte chains in solution

10.1 Introduction

Many macromolecules have ionizable moieties that can lead to charged species in aqueous solutions. The structure of linear polyelectrolyte molecules in dilute solution is a very strong function of the ionic strength of the solution and the charge density along the chain. The character of these polymers is treated in Section 10.2.

The thermodynamics of polyelectrolyte solutions is also a strong function of ionic strength and charge density. Very large osmotic second virial coefficients are observed for highly charged macromolecules. The Flory theory of these solutions in presented in Section 10.3.

The viscosity of polyelectrolyte solutions varies dramatically with ionic strength and pH. The theory and experimental situation for these solutions is outlined in Section 10.4

10.2 Structure of linear polyelectrolyte chains in dilute solution

Consider a linear polyelectrolyte chain with m monomer units. Each monomer unit contains a chemical moiety that is sensitive to the aqueous environment. Examples of such groups include carboxylic acids, amine bases, or generalized bases with the corresponding counterions such as sulfate and sodium ions. For convenience, we will restrict attention to monovalent systems, but more complicated polymers such as DNA can be treated in the same manner. If the chain in dilute aqueous solution is electrically neutral, the conformation will be determined by the persistence length of the chain, the overall chain length, and the solvent quality, just as for any linear polymer in dilute solution. However, dissociation of some of the species in solution will increase the entropy substantially. Further dissociation will be resisted by electrostatic interactions between charged units along the chain. The chain will expand to minimize the charge repulsion, leading to a reduction in chain entropy. Eventually, these processes will reach equilibrium, and all the contributions to the chemical potentials will produce balance. The calculation of the equilibrium state of such a polyelectrolyte chain in dilute solution is one of the most challenging problems in polymer science.

One of the key concepts needed to carry out this calculation is the electrostatic potential, $\psi(\vec{r})$, in the solution due to the charges. For a single fixed charge of magnitude q_i at location $\vec{r}_i$ in a medium of dielectric constant D, the potential at location $\vec{r}$ can be expressed as:

$$\psi(\vec{r}) = \frac{q_i}{D|\vec{r} - \vec{r}_i|}. \tag{10.1}$$

For an actual ion characterized by a distance of closest approach, a, the potential near the ion in a salt solution of finite ionic strength is calculated to be:

$$\psi(r) = \frac{q_i}{D} \frac{\exp(\kappa a)}{1 + \kappa a} \frac{\exp(-\kappa r)}{r} \tag{10.2}$$

where r is the scalar distance from the center of the ion and the inverse Debye length is given by:

$$\kappa = \left(\frac{8\pi N_A e^2}{1000 D k_b T} \right)^{1/2} I^{1/2} \tag{10.3}$$

where e is the electron charge and I is the ionic strength.

The ionic strength depends on all the ions in solution. When κ is very small, the potential approaches the pure Coulomb result (Equation 10.1). When the ionic strength is large, or when the dielectric constant is small, the charge is screened by the solution when $\kappa r \gg 1$.

For a linear polyelectrolyte chain with Z net charges along its contour, the electrostatic energy of a chain of effective radius R_e has been reported by Tanford[40] to be:

$$W_{\text{el}} = \frac{3Z^2 e^2}{2DR_e} \left\{ \frac{1}{\kappa^2 R_e^2} - \frac{3}{2\kappa^5 R_e^5} \left[\kappa^2 R_e^2 - 1 + \left(1 + \kappa R_e \right)^2 \exp\left(-2\kappa R_e \right) \right] \right\}. \tag{10.4}$$

As the chain expands, the entropy is reduced, and the eventual state of balance is achieved by jointly minimizing the total free energy for the chain, as in the excluded-volume problem in Chapter 5.

Another approach is to enumerate the charged sites explicitly. The electrostatic energy is then calculated as a function of the end-to-end distance of the chain for a stretched random coil. The probability distribution for the end-to-end distance can then be expressed as:

$$\wp(R) = \frac{\wp_0(R)\exp(-W_{el}/k_bT)}{\int\limits_{all\ R}\wp_0(R)\exp(-W_{el}/k_bT)dR}. \tag{10.5}$$

For a set of Z discrete charges along a flexible chain with unperturbed distribution of end-to-end lengths given as in Equation 4.9, the electrostatic energy can be expressed as:

$$W_{el} = \left(\frac{Z^2e^2}{2D}\right)\left\langle\frac{\exp(-\kappa R_{ij})}{R_{ij}}\right\rangle \tag{10.6}$$

where the brackets denote an average over all configurations for pairs of charges separated by distance R_{ij}. When the average is carried out for a Gaussian chain, the result is:

$$W_{el}(R) = \left(\frac{Z^2e^2}{DR}\right)\ln\left[1+\left(\frac{6R}{\kappa\langle R^2\rangle_0}\right)\right]. \tag{10.7}$$

The balance between electrostatic expansion and entropic resistance can be carried out as a force balance. The entropic retractive force due to chain extension is given by Morawetz[29] as:

$$f(R) = \frac{k_bT}{\langle R^2\rangle_0^{1/2}}\left[\frac{3R}{\langle R^2\rangle_0^{1/2}} - \frac{2\langle R^2\rangle_0^{1/2}}{R}\right]. \tag{10.8}$$

When the end-to-end distance has its most probable value for a Gaussian chain, $R_{mp} = (2\langle R^2\rangle_0/3)^{1/2}$, the retractive force is zero.

If the chain extension approaches its maximum value, it is possible to use a more precise expression for the distribution of end-to-end distances as derived by Kuhn et al.:[25]

$$\wp_0\left(R\right)=\left(\frac{3}{2\pi\left\langle R^2\right\rangle_0^{1/2}}\right)^{3/2}\left[\exp\left(-\frac{\int_0^R L^{-1}\left(R/R_{max}\right)dR}{R_{max}/m}\right)\right]4\pi R^2 \quad (10.9)$$

where the inverse Langevin function is a function of the relative extension. For this limit, the force of retraction is:

$$f\left(R\right)=\frac{k_bT}{\left\langle R^2\right\rangle_0^{1/2}}\left[\frac{R_{max}L^{-1}\left(R/R_{max}\right)}{\left\langle R^2\right\rangle_0^{1/2}}-\frac{2\left\langle R^2\right\rangle_0^{1/2}}{R}\right]. \quad (10.10)$$

The force due to the electrostatic repulsion is obtained from $f_{el}=\left(\partial E_{el}/\partial R\right)_T$. The force balance then yields the condition of equilibrium for the most probable value of the end-to-end length at which the net force is zero:

$$\frac{R_{mp}^2}{\left\langle R^2\right\rangle_0}-\frac{2}{3}=\left(\frac{Z^2e^2}{3Dk_bTR_{mp}}\right)\left[\ln\left(1+\frac{6R_{mp}}{\kappa\left\langle R^2\right\rangle_0}\right)-\frac{\left(6R_{mp}/\kappa\left\langle R^2\right\rangle_0\right)}{1+\left(6R_{mp}/\kappa\left\langle R^2\right\rangle_0\right)}\right]. \quad (10.11)$$

While there are many subtle issues that must be included to improve on the model outlined above, the value of the net charge on the chain is the most important. In a dilute solution with no added salt, the Debye length can be very large. If the Debye length is comparable to the size of the unperturbed radius of gyration, all the charges inside the coil will be essentially unscreened. Since the fields inside the coil will be enormous, very few of the counterions will venture outside the pervaded volume of the chain and will serve to screen the interactions within the macromolecule. The relevant value of the inverse Debye length will then be determined by the ionic strength inside the coil rather than the mean value for the solution. That chains will expand when the net charge is increased has been demonstrated for poly(methacrylic acid). The pure polymer contains weak acid moieties along the chain and, in pure water, is only partially ionized. Titration with sodium hydroxide increases the net charge on the chain, and the chain is observed to expand in Figure 10.1. Suggestions that fully titrated chains would be fully extended are unlikely. The data levels off for high degrees of neutralization, since the sodium ions serve to screen the carboxylate anions produced by titration.

The counterions are distributed both within the pervaded volume of the chain and in the free solution, with a distribution determined by the potential of the polyelectrolyte and its entrained counterions. The free counterions

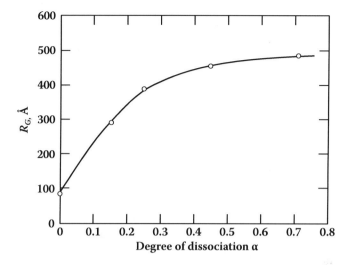

Figure 10.1 The measured radius of gyration for poly(methacrylic acid) obtained from light scattering as a function of the degree of acid dissociation produced by titration with base. The chain expands and reaches a limiting expansion. (From Tanford, C., *Physical Chemistry of Macromolecules*, John Wiley and Sons, New York, 1961. With permission.)

will not be as visible to the light scattering as the coil, but the concentration should fall off proportional to $\exp(-e\psi/k_bT)$. Evidence for the free counterions can be obtained from measurements of the osmotic pressure.

When the ionic strength increases to the point where the Debye length is comparable to the size of a statistical subunit, the polyelectrolyte problem can be approached in the same manner as the normal excluded-volume problem. The limiting form for Equation 10.11 becomes:

$$\frac{R_{mp}^2}{\langle R^2 \rangle_0} - \frac{2}{3} = \frac{3Z^2 R_{mp}}{2\pi DRTI \langle R^2 \rangle_0^2}. \qquad (10.12)$$

Another approach taken by Flory is to consider the polyelectrolyte coil under large ionic strength conditions as a phase subject to the conditions of the Donnan equilibrium. The coil will expand to reduce the Donnan potential, but the chain will resist due to the decreasing entropy. The coil is characterized by an effective volume, V_s, and the free salt concentration is c_s^0. The condition of equilibrium is given by Morawetz as:

$$\frac{R_{mp}^2}{\langle R^2 \rangle_0} - \frac{2}{3} = 2c_s^0 V_s \left(\sqrt{1 + \left(Z / 2c_s^0 V_s \right)^2} - 1 \right). \qquad (10.13)$$

The number of counterions due to the polyelectrolyte molecule within the volume of the coil is equal to Z for univalent systems. The concentration within the coil is then Z/V_s. A more sophisticated treatment by Flory includes the effect of solvent quality and the expansion or contraction caused by local interactions as well as the nonuniform distribution of charge within the coil.

10.3 Thermodynamics of polyelectrolyte solutions

The osmotic second virial coefficient for a polyelectrolyte solution depends on the potential of mean force for two chains. Since the effective components in such a system include the counterions needed to produce neutral species, it is a subtle combination of factors that determines the sign and value of A_2. The ionic strength of a polyelectrolyte solution increases with the concentration of polymer whether background salts are added or not. This fact adds another level of complexity to the treatment of the osmotic pressure of a polyelectrolyte solution as a function of concentration.

A comparison of the effect of concentration on the osmotic pressure of a polyelectrolyte is presented in Figure 10.2, along with the corresponding neutral chain in pure solvent and the polyelectrolyte chain in high salt. When the charge is either absent or fully screened, the osmotic pressure reflects the molecular weight of the chain and, in the solvents used in this study, displays a positive second virial coefficient. In pure alcohol, the osmotic pressure is much greater for the polyelectrolyte chain. The primary reason for the increase is the free counterions liberated into the solution. Since the osmotic pressure is produced by the kinetic motion of free solute particles, only a few of the counterions need to be free to produce a large increase in osmotic pressure. The decrease in π/c as the polyelectrolyte concentration is increased appears to be very strange, and only recently has some insight been gained into the reason for this behavior. In the absence of added salt, the second virial coefficient for polyelectrolyte chains is actually negative. There is a net attraction that allows the chains to share counterions. Thus the slope is negative. Three-body interactions are again positive, and the decrease is moderated. Evidence of the importance of these attractive forces has been obtained from light scattering.

The theory of the second osmotic virial coefficient in moderate salt solutions was developed in detail by Orofino and Flory.[30] An extension of the Flory–Krigbaum approach was also developed. The chemical potential of the solvent in a nonionic solution was expressed as:

$$\mu_1 - \mu_1^0 = -RT\left[\phi_2 / x + \left(1/2 - \chi_1\right)\phi_2^2 + \left(1/3 - \chi_2\right)\phi_2^3 + \cdots\right] \quad (10.14)$$

where x is the ratio of molar volume cited in Chapter 5, and a new Flory parameter is introduced for the interaction of three units. This approach leads to an extension of the Flory–Krigbaum potential:

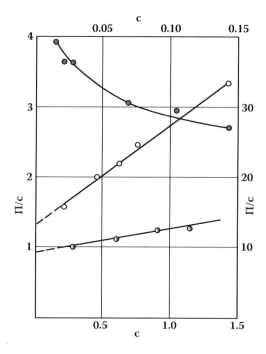

Figure 10.2 Comparison of the osmotic pressure ratio, π/c, for the neutral polymer, poly(4-vinylpyridine) in alcohol (o, left and below) with the derivative polyelectrolyte poly(N-butyl-4-vinylpyridinium bromide). Left and below, the polyelectrolyte is in alcoholic 0.61 N LiBr solution, a high-ionic-strength system. The lower intercept is due to the increase in molecular weight with derivatization. Right and above, the polyelectrolyte chain is in pure alcohol. The increase at low concentrations is dramatic. (From Flory, P.J., *Principles of Polymer Chemistry,* Cornell University Press, Ithaca, NY, 1953. With permission.)

$$U\left(R_{12}\right) = k_b T \left[X_1 \exp\left(-3R_{12}^2 / 4\left\langle R_G^2 \right\rangle\right) + X_2 \exp\left(-R_{12}^2 / \left\langle R_G^2 \right\rangle\right) \right] \quad (10.14)$$

where:

$$X_1 = \left(1/4\right)\left(3/\pi\right)^{3/2}\left(\overline{v}_2^2 / V_1 N_A\right)\left(M^2 / \left\langle R_G^2 \right\rangle^{3/2}\right)\left(1/2 - \chi_1\right)$$

$$X_2 = 3\left(3^{1/2}/2\pi\right)^3\left(\overline{v}_2^3 / V_1 N_A^2\right)\left(M^3 / \left\langle R_G^2 \right\rangle^3\right)\left(1/3 - \chi_2\right)$$

$\overline{v}_2$ = partial specific volume of the polymer

V_1 = molar volume of the solvent

The second virial coefficient is then obtained as in Equation 5.21. The result is:

$$A_2 =$$

$$\left(16\pi / 3^{3/2}\right)\left(N_A \left\langle R_G^2 \right\rangle^{3/2} / M^2 \right)\ln\left[1 + \left(\pi^{1/2} / 4\right)X_1 + \left(\pi^{1/2} 3^{3/2} / 32\right)X_2\right]. \quad (10.17)$$

It is important to note that, in this treatment, it is necessary for X_1 to be negative to balance the effect of three mer interactions if the second virial coefficient is to be zero.

   Inclusion of the Donnan effect in the calculation of the intermolecular potential produces the same form with new definitions for the coefficients. For the univalent case considered here, X_2 remains the same. A new contribution to the last factor in X_1 due to the Donnan potential appears as:

$$\left(1/2 - \chi_1 + V_1 \left(Z / m\right)^2 / 4 \left(V_2 / m\right)^2 I\right). \quad (10.18)$$

The second osmotic virial coefficient for a polyelectrolyte in a salt solution is then given approximately by Equation 10.17, with the new expression for X_1. The value of the second virial coefficient is raised by the inclusion of the Donnan effect.

10.4   Viscosity of polyelectrolyte solutions

The expansion of linear polyelectrolyte chains in pure water is demonstrated even more dramatically in the viscosity of such solutions. The intrinsic viscosity depends on the expansion coefficient to the third power, so that an increase in the mean-squared end-to-end distance of a factor of 2 would lead to an increase of a factor of 8 in the intrinsic viscosity. The behavior of the reduced specific viscosity as a function of polymer concentration under different ionic strength conditions is shown in Figure 10.3. For high ionic strength, the reduced specific viscosity, $\eta_{sp} / c = \left(\eta - \eta_0\right) / \eta_0 c$, depends weakly on polymer concentration. In pure water, the reduced specific viscosity becomes very large as the concentration is reduced. At intermediate ionic strength, the background electrolyte hinders expansion of the chains as the concentration of counterions due to the polymer is reduced below the background level. Empirically, the reduced specific viscosity is described by:

$$\eta_{sp} / c = \frac{A}{1 + B\sqrt{c}}. \quad (10.19)$$

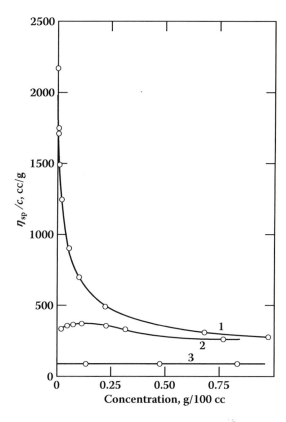

Figure 10.3 Reduced specific viscosity of poly(vinylbutylpyridinium bromide) in pure water (1), 0.001*M* KBr (2), and 0.0335*M* KBr (3) as a function of polymer concentration. (From Tanford, C., *Physical Chemistry of Macromolecules*, John Wiley and Sons, New York, 1961. With permission.)

References

1. Akcasu, A.Z., Benmouna, M., and Han, C.C., Interpretation of dynamic scattering from polymer solution, *Polymer*, 21, 866, 1980.
2. Alms, G.R., Patterson, G.D., and Stevens, J.R., A study of the thermal polymerization of styrene by depolarized Rayleigh scattering spectroscopy, *J. Chem. Phys.*, 70, 2145, 1979.
3. Berne, B.J. and Pecora, R., *Dynamic Light Scattering*, Wiley-Interscience, New York, 1976.
4. Bovey, F.A., *Chain Structure and Conformation of Macromolecules*, Academic Press, New York, 1982.
5. Carraway, R. and Leeman, S.E., The amino acid sequence of a hypothalamic peptide, neurotensin, *J. Biol. Chem.*, 250, 1907, 1975.
6. Carroll, P.J. and Patterson, G.D., The distribution of relaxation frequencies from photon correlation spectroscopy near the glass transition, *J. Chem. Phys.*, 82, 9, 1985.
7. Cassasa, E.F., Phase Equilibrium in Polymer Solutions, in *Fractionation of Synthetic Polymers*, Tung, L.H., Ed., Marcell Dekker, New York, 1977, pp. 1–102.
8. Croft, L.R., *Handbook of Protein Sequence Analysis*, John Wiley & Sons, New York, 1980.
9. deGennes, P.-G., *Scaling Concepts in Polymer Physics*, Cornell University Press, Ithaca, NY, 1979.
10. Doolittle, A.K., Studies in Newtonian Flow II. Dependence of the viscosity of liquids on free space, *J. Appl. Phys.*, 22, 1471, 1951.
11. Ediger, M.D., Heterogeneous dynamics in superspooled liquids, *Annu. Rev. Phys. Chem.*, 51, 99, 2000.
12. Ferry, J.D., *Viscoelastic Properties of Polymers*, John Wiley & Sons, New York, 3rd ed., 1980.
13. Flory, P.J. and Bueche, A.M., Theory of light scattering by polymer solutions, *J. Polym. Sci.*, 27, 219, 1958.
14. Flory, P.J. and Krigbaum, W.R., Statistical mechanics of dilute polymer solutions, II, *J. Chem. Phys.*, 18, 1086, 1950.
15. Flory, P.J., *Principles of Polymer Chemistry*, Cornell University Press, Ithaca, NY, 1953.
16. Flory, P.J. and Osterheld, J.E., Intrinsic viscosities of polyelectrolytes: polyacrylic acid, *J. Phys. Chem.*, 58, 653, 1954.

17. Flory, P.J., Phase equilibria in solutions of rod-like particles, *Proc. R. Soc., A,* 234, 73, 1956.
18. Flory, P.J., Statistical thermodynamics of liquid mixtures, *J. Am. Chem. Soc.,* 87, 1833, 1965.
19. Flory, P.J., *Statistical Mechanics of Chain Molecules,* Interscience, New York, 1969.
20. Francis, R.S., Patterson, G.D., and Kim, S.H., Liquid-like structure of polymer solutions near the overlap concentration, *J. Polym. Sci. B: Polym. Phys.,* 44, 703, 2006.
21. Huggins, M. L., *Physical Chemistry of High Polymers,* John Wiley and Sons, New York, 1958.
22. Huglin, M.B., Ed., *Light Scattering from Polymer Solutions,* Academic Press, New York, 1972.
23. Kim, S.H., Ramsay, D.J., Patterson, G.D., and Selser, J.C., Static and dynamic light scattering of poly(α-methyl styrene) in toluene in the dilute region, *J. Polym. Sci. B: Polym. Phys.,* 28, 2023, 1990.
24. Kovacs, A.J., Glass transition in amorphous polymers: a phenomenological study, *Adv. Polym. Sci.,* 3, 394, 1963.
25. Kuhn, Kunzle, and Katchalsky, The behavior of multivalent chain molecules in solution, *Helvetica Chimica Acta,* 31, 1994, 1948.
26. Lindsey, C.P. and Patterson, G.D., Detailed comparison of the Williams-Watts and Cole-Davidson functions, *J. Chem. Phys.,* 73, 3348, 1980.
27. McCrum, N.G., Read, B.E., and Williams, G., *Anelastic and Dielectric Effects in Polymeric Solids,* John Wiley & Sons, New York, 1967.
28. Min, G. et al., Solution characterization of monodisperse atactic polystyrenes by static and dynamic light scattering, *Int. J. Polym. Anal. Characterization,* 8, 187, 2003.
29. H. Morawetz, Theoretical Aspects of Chain Configuration and Counterion Distribution in Solutions of Flexible Chain Polyelectrolytes, in *Polyelectrolyte Solutions,* Rice, S.A., Nagasawa, M., Eds., Academic Press, New York, 1961.
30. Morawetz, H., *Polymers: the Origins and Growth of a Science,* John Wiley & Sons, New York, 1985.
31. Nose, T. and Chu, B., Static and dynamical properties of polystyrene in transdecalin III: Polymer dimensions in dilute solution in the transition region, *Macromolecules,* 12, 1122, 1979.
32. Orofino, T.A. and Flory, P.J., The second virial coefficient for polyelectrolytes, *J. Phys. Chem.,* 63, 283, 1959.
33. Orwoll, R.A. and Flory, P.J., Thermodynamic properties of binary mixtures of n-alkanes, *J. Am. Chem. Soc.,* 89, 6814, 1967.
34. Patterson, G.D., Dynamic light scattering from bulk polymers, *Ann. Rev. Mater. Sci.,* 13, 219, 1984.
35. Patterson, G.D., Ramsay, D.J., and Carroll, P.J., Depolarized light scattering spectroscopy and polymer characterization, *Analytica Chimica Acta,* 189, 57, 1986.
36. Perrin, F., Brownian movement of an ellipsoid. II. Free rotation and depolarization of fluorescence. Translation and diffusion of ellipsoidal molecules, *J. Phys. Radium,* 7, 1, 1936.
37. Savin, D.A., Patterson, G.D., and Stevens, J.R., Evidence for the γ-relaxation in the light scattering spectra of poly(n-hexyl methacrylate) near the glass transition, *J. Polym. Sci. B: Polym. Phys.,* 43, 1504, 2005.

38. Snyder, R.G., Infrared and raman spectra of polymers, in *Methods of Experimental Physics*, Fava, R.A., Ed., Academic Press, New York, 1980, vol. 16A, pp. 73–148.
39. Struick, L.C.E., *Physical Aging in Amorphous Polymers and Other Materials*, Elsevier, Amsterdam, 1978.
40. Tanford, C., *Physical Chemistry of Macromolecules*, John Wiley & Sons, New York, 1961.
41. Thomas, E.L. and Lescanec, R.L., Phase morphology in block copolymer systems, *Phil. Trans. R. London A*, 348, 149, 1994.
42. Treloar, L.R.G., *The Physics of Rubber Elasticity*, Clarendon Press, Oxford, U.K., 1975.
43. Wong, P.T.T., Mantsch, H.M., and Snyder, R.G., *J. Chem. Phys.*, 47, 1316, 1967.
44. Yamakawa, H., Effects of pressure on conformer equilibria in liquid n-hexane, *Modern Theory of Polymer Solutions*, Harper and Row, New York, 1971.

Index